Lecture Notes in Computer Science 4893

Commenced Publication in 1973
Founding and Former Series Editors:
Gerhard Goos, Juris Hartmanis, and Jan van Leeuwen

T0223134

Solomon W. Golomb   Guang Gong
Tor Helleseth   Hong-Yeop Song (Eds.)

Sequences,
Subsequences,
and Consequences

International Workshop, SSC 2007
Los Angeles, CA, USA, May 31 - June 2, 2007
Revised Invited Papers

 Springer

Volume Editors

Solomon W. Golomb
University of Southern California, Department of EE-Systems
Los Angeles, CA 90089-2565, USA
E-mail: sgolomb@usc.edu

Guang Gong
University of Waterloo, Department of Electrical and Computer Engineering
200 University Avenue West, Waterloo, ON N2L 3G1, Canada
E-mail: g.gong@calliope.uwaterloo.ca

Tor Helleseth
University of Bergen, Selmer Center, Department of Informatics
Thormohlensgate 55, 5020 Bergen, Norway
E-mail: tor.helleseth@ii.uib.no

Hong-Yeop Song
Yonsei University, School of Electrical and Electronic Engineering
134 Sinchon-dong Seodaemun-gu, Seoul, 120-749, Korea
E-mail: hysong@yonsei.ac.kr

Library of Congress Control Number: 2007941501

CR Subject Classification (1998): E.3-4, G.2, F.2.2, C.2, J.1

LNCS Sublibrary: SL 4 – Security and Cryptology

ISSN   0302-9743
ISBN   3-540-77403-3 Springer Berlin Heidelberg New York
ISBN   978-3-540-77403-7 Springer Berlin Heidelberg New York

Springer is a part of Springer Science+Business Media

springer.com

© Springer-Verlag Berlin Heidelberg 2007

Typesetting: Camera-ready by author, data conversion by Scientific Publishing Services, Chennai, India
Printed on acid-free paper   SPIN: 12208166   06/3180   5 4 3 2 1 0

Dedicated to
Professor Solomon W. Golomb
on his 75th Birthday

Preface

These are the proceedings of the Workshop on Sequences, Subsequences, and Consequences that was held at the University of Southern California (USC), May 31 - June 2, 2007. There were three one-hour Keynote lectures, 16 invited talks of up to 45 minutes each, and 1 "contributed" paper.

The theory of sequences from discrete symbol alphabets has found practical applications in many areas of coded communications and in cryptography, including: signal patterns for use in radar and sonar; spectral spreading sequences for CDMA wireless telephony; key streams for direct sequence stream-cipher cryptography; and a variety of forward-error-correcting codes.

The workshop was designed to bring leading researchers on "sequences" from around the world to present their latest results, interchange information with one another, and especially to inform the larger audience of interested participants, including faculty, researchers, scholars, and students from numerous institutions, as well as the readers of these proceedings, about recent developments in this important field.

There were invited speakers from Canada, China, Germany, India, Israel, Norway, Puerto Rico, and South Korea, in addition to those from the USA. Support for the workshop was generously provided by the Office of the Dean of the Viterbi School of Engineering, by the Center for Communications Research (CCR-La Jolla), and by the United States National Science Foundation (NSF). This support is hereby gratefully acknowledged.

As the principal organizers of the workshop and its technical program, we wish to thank the speakers, all the participants, and the above-mentioned organizations that provided funding. Our special thanks go to Milly Montenegro, Gerrielyn Ramos, and Mayumi Thrasher at USC for all the arrangements before, during, and after the workshop, and for their considerable help, along with Xinxin Fan and Honggang Hu at the University of Waterloo (Canada) and Young-Joon Kim at Yonsei University (South Korea), in preparing, typing, and formatting many of the manuscripts.

September 2007

Solomon W. Golomb
Guang Gong
Tor Helleseth
Hong-Yeop Song

Organization

Technical Program Committee for SSC 2007

Principal Organizers

Solomon Golomb   University of Southern California, USA
Guang Gong   University of Waterloo, Canada
Tor Helleseth   University of Bergen, Norway
Hong-Yeop Song   Yonsei University, Korea

Program Committee

Tuvi Etzion   Technion IIT, Israel
Solomon W. Golomb   University of Southern California, USA
Guang Gong   University of Waterloo, Canada
Alfred W. Hales   CCR-La Jolla, USA
Tor Helleseth   University of Bergen, Norway
Alexander Kholosha   University of Bergen, Norway
Andrew Klapper   University of Kentucky, USA
P. Vijay Kumar   University of Southern California, USA, and
                            IISc Bangalore
Robert McEliece   California Institute of Technology, USA
Lothrop Mittenthal   Teledyne, USA
Oscar Moreno   University of Puerto Rico, Puerto Rico
Alexander Pott   Otto von Guericke University Magdeburg,
                            Germany
Hong-Yeop Song   Yonsei University, Korea
Xiaohu Tang   Southwest Jiao Tong University, China
Herbert Taylor   South Pasadena, USA
Andrew J. Viterbi   Viterbi Group, LLC, USA
Steven Wang   University of Carleton, Canada
Lloyd R. Welch   University of Southern California, USA
Nam Yul Yu   University of Waterloo, Canada

Sponsoring Institutions

University of Southern California, CCR-La Jolla, NSF

Table of Contents

Periodic Binary Sequences: Solved and Unsolved Problems

Solomon W. Golomb

University of Southern California

Abstract. The binary linear feedback shift register sequences of degree n and maximum period $p = 2^n - 1$ (the *m-sequences*) are useful in numerous applications because, although deterministic, they satisfy a number of interesting "randomness properties".

An important open question is whether a binary sequence of period $p = 2^n - 1$ with both the span-n property and the two-level correlation property must be an m-sequence.

There is a direct correspondence between m-sequences of degree n and primitive polynomials of degree n over $GF(2)$. Several conjectures are presented about primitive polynomials with a bounded number of terms.

1 Introduction

Feedback shift register sequences have been widely used as synchronization codes, masking or scrambling codes, and for white noise signals in communication systems, signal sets in CDMA (code division multiple access) communications, key stream generators in stream cipher cryptosystems, random number generators in many cryptographic primitive algorithms, and as testing vectors in hardware design.

Notation:

- $\mathbb{F} = GF(2) = \{0, 1\}$
- $\mathbb{F}_2^n = \{(a_0, a_1, \cdots, a_{n-1}) | a_i \in \mathbb{F}_2\}$, a vector space over $\mathbb{F}_2$ of dimension n.
- A *boolean function of* n *variables*, i.e., $f : \mathbb{F}_2^n \to \mathbb{F}_2$, which can be represented as follows:

$$f(x_0, x_1, \cdots, x_{n-1}) = \sum c_{i_1 i_2 \cdots i_t} x_{i_1} x_{i_2} \cdots x_{i_t}, c_{i_1 i_2 \cdots i_t} \in \mathbb{F} \qquad (1)$$

where the sum runs through all subsets $\{i_1, \cdots, i_t\}$ of $\{0, 1, \cdots, n-1\}$. This shows that there are 2^{2^n} different boolean functions of n variables.

An n-**stage shift register** is a circuit consisting of n consecutive 2-state storage units (flips-flops) regulated by a single clock. At each clock pulse, the state (1 or 0) of each memory stage is shifted to the next stage in line. A shift

S.W. Golomb et al. (Eds.): SSC 2007, LNCS 4893, pp. 1–8, 2007.

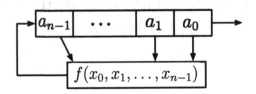

Fig. 1. A Diagram of FSR of Degree n

register is converted into a code generator by including a feedback loop, which computes a new term for the left-most stage, based on the n previous terms. In Figure 1, we see a diagram of a feedback shift register (FSR).

Each of the squares is a 2-state storage unit. The n binary storage elements are called the *stages* of the shift register, and their contents (regarded as either a binary number or a binary vector, n bits in length) is called a *state* of the shift register. $(a_0, a_1, \cdots, a_{n-1}) \in F^n$ is called an *initial state of the shift register*. The feedback function $f(x_0, x_1, \cdots, x_{n-1})$ is a boolean function of n variables, defined in (1). At every clock pulse, there is a transition from one state to the next. To obtain a new value for stage n, we compute $f(x_0, x_1, \cdots, x_{n-1})$ of all the present terms in the shift register and use this in stage n. For example, the next state of the shift register in Figure 1 becomes $(a_1, a_2, \cdots, a_n)$ where

$$a_n = f(a_0, a_1, \cdots, a_{n-1}).$$

After the consecutive clock pulses, a feedback shift register outputs a sequence:

$$a_0, a_1, \cdots, a_n, \cdots. \tag{2}$$

The sequence satisfies the following recursion relation

$$a_{k+n} = f(a_k, a_{k+1}, \cdots, a_{k+n-1}), k = 0, 1, \cdots. \tag{3}$$

Any n consecutive terms of the sequence in (2),

$$a_k, a_{k+1}, \cdots, a_{k+n-1}$$

represents a state of the shift register in Figure 1. A *state (or vector) diagram* is a diagram that is drawn based on the successors of each of the states. The output sequence is called a *feedback shift register sequence*.

If the feedback function $f(x_0, x_1, \cdots, x_{n-1})$ is a linear function, then the output sequence is called a *linear feedback shift register (LFSR) sequence*. Otherwise, it is called a *nonlinear feedback shift register (NLFSR) sequence*.

Examples. In Figure 2, we see a 3-stage shift register with a linear feedback function $f(x_0, x_1, x_2) = x_0 + x_1$.

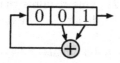

Fig. 2. An LFSR of Degree 3

In Figure 3, we see an n-stage shift register with a linear feedback function.

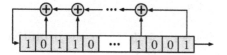

Fig. 3. An LFSR of Degree n

When f is linear, i.e.,

$$f(x_0, x_1, \cdots, x_{n-1}) = \sum_{i=0}^{n-1} c_i x_i, c_i \in \mathbb{F}_2$$

the elements of a satisfy the following recursion relation:

$$a_{k+n} = \sum_{i=0}^{n-1} c_i a_{k+1}, k = 0, 1, \cdots. \tag{4}$$

m-**Sequences:** $f(x) = \sum_{i=0}^{n-1} c_i x_i$ is primitive over $\mathbb{F}_2$.

2   Polynomial Conjectures

The shift register

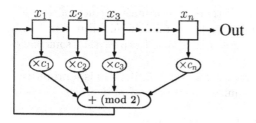

corresponds to the polynomial $f(x) = 1 + \sum_{i=1}^{n} c_i x^i$, and produces an m-sequence (starting from any initial state except $x_1 = x_2 = \cdots = x_n = 0$) if and only if $f(x)$ is **primitive** over $GF(2)$. This requires that $f(x)$ is **irreducible** over $GF(2)$, and is equivalent to: $f(x)$ divides $x^t + 1$ for $t = p = 2^n - 1$ and for no smaller positive integer value of t.

The number of such primitive polynomials of degree n over $GF(2)$ is known to be $\phi(2^n - 1)/n$, where ϕ is Euler's phi-function. This is the number of cyclically distinct m-sequences of degree n.

Many unsolved problems concern the existence of such primitive polynomials with a restricted number of terms. The two strongest conjectures are:

Conjecture 1. For infinitely many values of n there are **primitive trinomials**, $f(x) = x^n + x^a + 1, 0 < a < n$.

Conjecture 2. For all degrees $n \geq 5$, there are **primitive pentanomials**, $x^n + x^{a_3} + x^{a_2} + x^{a_1} + 1, 0 < a_1 < a_2 < a_3 < n$.

Perhaps the weakest conjecture of this type is:

Conjecture 3. There is a finite positive integer m such that, for infinitely many degrees n, there is a primitive polynomial $f(x)$ of degree n having no more than m terms.

Conjecture 1 is the special case of Conjecture 3 with $m = 3$. Conjecture 2 would imply Conjecture 3 with $m = 5$. I would hope that someone will prove Conjecture 3, perhaps with a very large value of m. Subsequent effort would then be devoted to reducing this value of m.

Even stronger than Conjecture 1 would be:

Conjecture 4. The trinomial $x^n + x + 1$ is primitive for infinitely many values of n.

For the foreseeable future I do not expect this to be proved.

An empirical observation that I made some 50 years ago, that there are no primitive trinomials when n is a **multiple of 8**, was proved by Richard Swan, who showed that $x^n + x^a + 1$, $0 < a < n$ and n a multiple of 8, has an **even number** of irreducible factors.

I can easily prove: If there is a primitive trinomial of odd degree $n \geq 5$, then there is also a primitive pentanomial of this degree n. I conjecture that this is also true for even degree $n > 5$, and I expect that someone can prove this, as:

Conjecture 5. For all degrees $n \geq 5$, if there is a primitive trinomial of degree n, then there is a primitive pentanomial of degree n.

Over 75 years ago, Øystein Ore proved:

Theorem1. If $f(x) = 1 + \sum_{i=1}^{n} c_i x^i$ is primitive, then $f(x) = 1 + \sum_{i=1}^{n} c_i x^{2^i - 1}$ is irreducible.

3   "Randomness Properties" of m-Sequences

It is the remarkable "randomness properties" of the m-sequences that makes them so useful in many applications: **cryptography, radar, GPS, "Monte Carlo" random number generation, CDMA, etc.**

Here are the "randomness properties" of these m-sequences (the ones with maximum period $2^n - 1$.

P-1. In each period of $2^n - 1$ bits, there are 2^{n-1} 1's and $2^{n-1} - 1$ 0's. (The **balance property**).

P-2. (The **"Run Property"**.) In each period there are 2^{n-1} "runs". Since runs of 1's alternate with runs of 0's, half the runs (2^{n-2}) are runs of 1's, and half (2^{n-2}) are runs of 0's. Half the runs of each type have length 1; $\frac{1}{4}$ of the runs of each type have length 2; $\frac{1}{8}$ of the runs of each type have length 3; $\cdots$; there is **one** run of each type of length $n - 2$; finally, there is **one** run of 0's of length $n - 1$, and **one** run of 1's of length n. (This is the expected distribution of run lengths when tossing a **random perfect coin**.)

P-3. (The **"Span n" Property**.) As we slide a window of length n around one cycle of the sequence, we see every n-bit binary number, except $00\cdots0$, exactly once.

P-4. (The **Multiplier Property**.) If we take every second term of the sequence (repeating around the cycle to get a full $2^n - 1$ terms), we get back the same sequence (possibly rotated cyclically).

P-5. (The **2-level autocorrelation property**.) If we compare the sequence of period $p = 2^n - 1$ with each of its (non-zero) cyclic shifts, we see 2^{n-1} **disagreements** and $2^{n-1} - 1$ **agreements**.

We define the **"autocorrelation function"** $C(\tau)$ of the sequence, at a shift of τ, to be $C(\tau) = \frac{A_\tau - D_\tau}{A_\tau + D_\tau}$, where $A_\tau = \#$ of agreements at a shift of τ, and $D_\tau = \#$ of disagreements at a shift of τ. We find

$$C(\tau) = \begin{cases} 1 & \text{at } \tau = 0 \\ -\frac{1}{p} & \text{for } 0 < \tau < p. \end{cases}$$

Example

τ	shifted sequence	A_τ	D_τ	$C(\tau)$
0	1 1 1 0 1 0 0	7	0	1
1	0 1 1 1 0 1 0	3	4	$-1/7$
2	0 0 1 1 1 0 1	3	4	$-1/7$
3	1 0 0 1 1 1 0	3	4	$-1/7$
4	0 1 0 0 1 1 1	3	4	$-1/7$
5	1 0 1 0 0 1 1	3	4	$-1/7$
6	1 1 0 1 0 0 1	3	4	$-1/7$

It is this **autocorrelation property** that makes these sequences ideal for use in **radar**.

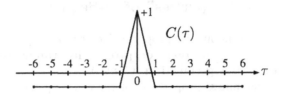

P-6. The **Cycle-and-Add Property**: Every m-sequence has the property that, when added, term-by-term "modulo 2", to any cyclic shift of itself, what results is merely a new cyclic shift.

Example

$$
\begin{array}{cccc}
1110100 & 1110100 & 1110100 & 1110100 \\
+\,0111010 & +\,0011101 & +\,1001110 & +\,0100111 \\
\hline
1001110 & 1101001 & 0111010 & 1010011
\end{array}
$$

3.1 Relationships of These "Randomness Properties" (Among All Binary Sequences with Period $2^n - 1$)

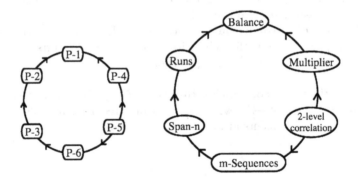

- There are $\dbinom{2^n-1}{2^{n-1}} = \frac{(2^n-1)!}{2^{n-1}!(2^{n-1}-1)!}$ **P-1** sequences

- There are $2^{2^{n-1}-n} = \frac{2^{2^{n-1}}}{2^n}$ **P-3** sequences.

- There are only $\frac{\phi(2^n-1)}{n} \le \frac{2^n-2}{n}$ **P-6** "m-sequences".

3.2 My Main Conjecture

If a sequence satisfies both **P-3** and **P-5**, it **must** be an m-sequence (**P-6**)!

Conjecture 6. A sequence of period $2^n - 1$ which has both span n and two-level autocorrelation must be an m-sequence.

An example of a sequence with both **P-2** (the run property) and **P-4** (the multiplier property) which has neither **P-3** nor **P-5** already occurs at period $p = 2^5 - 1 = 31$.

Unsolved Problem: For what larger degrees n do such examples occur?

An example of a sequence with both **P-3** (the span n property) and **P-4** (the multiplier property), but without **P-5** (the 2-level correlation property) occurs at period $p = 2^7 - 1 = 127$.

Such an example will occur whenever $x^n + x + 1$ is primitive, with odd $n > 3$ (as here with $n = 7$).

Unsolved Problem: Are these the only examples of sequences with both **P-3** and **P-4** but without **P-5** (and therefore not m-sequences)?

An example of a sequence with both **P-2** (the run property) and **P-5** (2-level autocorrelation) but without **P-3** (the span n property) occurs at period $p = 2^7 - 1 = 127$. (This example was found in an example of a sequence in one of the six families of **P-5** sequences in Baumert's book.)

Unsolved Problem: What other examples are there of sequences with **P-2** and **P-5** but without **P-3** (and therefore not m-sequences)?

As early as 1954, I asked the question: **What are all the balanced binary sequences with 2-level autocorrelation?**

At that time, I knew only of the m-sequences and the Quadratic Residue (Legendre) sequences. This question can be formulated as: **"What are all the cyclic Hadamard difference sets?"**

These are the cyclic (v, k, λ) difference sets with $v = 4t - 1, k = 2t - 1$, and $\lambda = t - 1$.

Conjecture 7. For such a difference set to exist, v **must have one of three forms:**

i) $v = 4t - 1$ prime;
ii) $v = 4t - 1 = p(p + 2)$, a product of the twin primes p and $q = p + 2$; and
iii) $v = 4t - 1 = 2^n - 1$.

A stronger conjecture is:

Conjecture 8. The only cyclic (v, k, λ) Hadamard difference sets when $\mathbf{v \neq 2^n - 1}$ come from three constructions:

i) $v = 4t - 1$ prime with the **Quadratic Residue construction;**
ii) $v = 4t - 1 = 4a^2 + 27$ prime, with **Hall's sextic residue sequence construction**; and
iii) $v = 4t - 1 = p(p + 2)$, p and $p + 2$ are both primes, with the **twin prime construction** (using the Jacobi symbol) of Stanton and Sprott.

Extensive computer searches have failed to find any other examples, with $v \neq 2^n - 1$.

3.3   Cyclic Hadamard Difference Sets when $v = 2^n - 1$

Finally, we have the question of cyclic Hadamard difference sets when $v = 2^n - 1$. The **known** examples include:

  i) the m-sequences, for all $n \geq 2$ (Singer difference sets);
 ii) the **Quadratic Residue sequences** when $v = 2^n - 1$ is a Mersenne prime;
iii) the **Hall's sextic residue sequences** when $v = 2^n - 1 = 4a^2 + 27 =$ prime, where it has been shown that there are only three such cases ($v = 2^5 - 1 = 31$, $v = 2^7 - 1 = 127$, and $v = 2^{17} - 1 = 131,071$);
 iv) the **GMW construction and all its generalization**, when n (in $v = 2^n - 1$) is composite, $n > 4$;
  v) **3-term and 5-term** sums of decimations of m-sequences, and the **Kasami power function constructions**;
 vi) **the Welch-Gong transforms** of 5-term sequences;
vii) the three families of **"hyper-oval sequences"**: Segré sequences, and Glynn type 1 and type 2 sequences.

Details of all these constructions, including variant constructions that lead to the same sequences, can be found in the 2005 book **Signal Design for Good Correlation**, by S. Golomb and G. Gong.

Exhaustive computer searches have been completed for all cyclic Hadamard difference sets of period $2^n - 1$ for all $n \leq 10$. (Starting at $n = 5$ in the early 1950's, one new value of n has been successfully searched in each succeeding decade.)

The aforementioned constructions account for all of the individual examples which have been found. Hence:

Conjecture 9. All the constructions which yield **cyclic Hadamard difference sets** are now known.

I am not sure that I believe this one.

References

1. Golomb, S.: Shift Register Sequences. Holden-Day Inc., San Francisco (1967); revised edition Aegean Park Press, Laguna Hills, CA (1982)
2. Golomb, S., Gong, G.: Signal Design for Good Correlation – for Wireless Communication, Cryptography, and Radar. Cambridge University Press, Cambridge (2005)

On Boolean Functions Which Are Bent and Negabent

Matthew G. Parker[1] and Alexander Pott[2]

[1] The Selmer Center, Department of Informatics, University of Bergen, N-5020 Bergen, Norway
[2] Institute for Algebra and Geometry, Faculty of Mathematics, Otto-von-Guericke-University Magdeburg, D-39016 Magdeburg, Germany

Abstract. Bent functions $f : \mathbb{F}_2^m \to \mathbb{F}_2$ achieve largest distance to all linear functions. Equivalently, their spectrum with respect to the Hadamard-Walsh transform is flat (i.e. all spectral values have the same absolute value). That is equivalent to saying that the function f has optimum periodic autocorrelation properties. Negaperiodic correlation properties of f are related to another unitary transform called the nega-Hadamard transform. A function is called *negabent* if the spectrum under the nega-Hadamard transform is flat. In this paper, we consider functions f which are simultaneously bent and negabent, i.e. which have optimum periodic and negaperiodic properties. Several constructions and classifications are presented.

Keywords: bent function, Boolean function, unitary transform, Hadamard-Walsh transform, correlation.

1 Introduction

Boolean functions $f : \mathbb{F}_2^m \to \mathbb{F}_2$ play an important role in cryptography. They should satisfy several properties, which are quite often impossible to be satisfied simultaneously. One property is the nonlinearity of a Boolean function, which means that the function is as far away from all linear functions as possible. Functions which achieve this goal are called *bent functions*. Equivalently, all Hadamard-Walsh coefficients of f are equal in absolute value.

There is another criteria which may be viewed as the negaperiodic analogue of the bent criteria. In spectral terms, it may be formulated as follows: Find functions whose nega-Hadamard spectrum is flat, i.e. all spectral values under the nega-Hadamard transform are equal in absolute value. Many bent functions are known, and also many negabent functions are known: It turns out that every linear function is negabent! In this paper, we are going to investigate the intersection of these two sets, i.e. we are searching for bent functions which are simultaneously negabent. At first view, it is not clear that such objects exist. An infinite series of bent-negabent functions has been found in [1,2].

We give necessary and sufficient conditions for quadratic functions to be both bent and negabent, which is based on [2]. It turns out that such quadratic *bent-negabent* functions exist for all even m, which generalizes the series in [2].

S.W. Golomb et al. (Eds.): SSC 2007, LNCS 4893, pp. 9–23, 2007.

More generally, we can describe all Maiorana-McFarland type bent functions which are simultaneously negabent. It seems to be difficult to exploit this condition in general.

The concept of a dual bent function is well known. If f is bent-negabent, then the dual has the same property. There is another interesting transformation which turns a bent-negabent function into a bent-negabent function. We call this *Schmidt complementation* since it is based on a construction in [4]. Therefore, we can construct orbits of bent-negabent functions starting from just one example. We may repeatedly apply dualization and Schmidt complementation. We will report some computational results.

This paper is organized as follows. In Section 2 we summarize some of the main results on bent and negabent functions which are needed throughout this paper.

In Section 3, we consider quadratic bent-negabent functions. In Section 4 we investigate Maiorana-McFarland bent functions. Transformations which preserve bent-negabentness are investigated in Section 5, in particular the Schmidt complementation. Finally, computational results are contained in the last Section 6.

2   Preliminaries

Let V_m denote the m-dimensional vector space $\mathbb{F}_2^m$. We consider functions $\tilde{f} : V_m \to \mathbb{C}$. In many cases, the image set is just $\{\pm 1\}$. Then we say that the function is *Boolean*. If $f : V_m \to \mathbb{F}_2$, we may easily turn it into a "complex-valued" Boolean function:

$$\tilde{f}(\mathbf{x}) := (-1)^{f(\mathbf{x})}.$$

Conversely, any function $\tilde{f} : V_m \to \{\pm 1\}$ determines a function $f : V_m \to \mathbb{F}_2$ by replacing -1 by 1 and 1 by 0. We also call f *Boolean*. The set of Boolean functions $\tilde{f} : V_m \to \mathbb{C}$ is embedded in a 2^m-dimensional unitary vector space $\mathcal{V}$ with an inner product

$$(\tilde{f}, \tilde{g}) = \sum_{\mathbf{x} \in V_m} \tilde{f}(\mathbf{x}) \overline{\tilde{g}(\mathbf{x})}. \tag{1}$$

A function $\tilde{f} : V_m \to \mathbb{C}$ is determined by the values $\tilde{f}(\mathbf{x})$. It will be useful to interpret this vector of "function values" as a polynomial in $\mathbb{C}[\xi_1, \ldots, \xi_m]$: We define the multivariate polynomial

$$F = \sum_{\mathbf{x} \in V_m} a_{\mathbf{x}} \xi^{\mathbf{x}}, \tag{2}$$

where $\xi^{\mathbf{x}} := \xi_1^{x_1} \cdots \xi_m^{x_m}$, and $a_{\mathbf{x}} = \tilde{f}(\mathbf{x})$ for $\mathbf{x} \in V_m$. We call F the *indicator polynomial* of $\tilde{f}$. If $f : V_m \to \mathbb{F}_2$, we first have to turn f into a complex-valued function $(-1)^f$, as described above.

Note that $f : V_m \to \mathbb{F}_2$ itself may also be defined as a multivariate polynomial. Both polynomials, f and its indicator F, describe the same object of interest (the Boolean function f), but in a completely different way. Therefore, we will write $\mathbf{x}$ when we deal with $f : V_m \to \mathbb{F}_2$, and ξ when dealing with the indicator.

The set of polynomials $\sum_{\mathbf{x} \in V_m} a_\mathbf{x} \xi^\mathbf{x}$ forms a complex vector space $\mathcal{L}$ of dimension 2^m. On this vector space, we define the usual inner product:

$$(F, G) := \sum_{\mathbf{x} \in V_m} a_\mathbf{x} \overline{b_\mathbf{x}},$$

where $F = \sum a_\mathbf{x} \xi^\mathbf{x}$, $G = \sum b_\mathbf{x} \xi^\mathbf{x}$. If F and G are the indicator polynomials of two functions $\tilde{f}$ and $\tilde{g}$, then

$$(F, G) := \sum_{\mathbf{x} \in V_m} \tilde{f}(\mathbf{x}) \overline{\tilde{g}(\mathbf{x})},$$

which is the same as (1). This shows that the indicator map $\mathcal{I} : \mathcal{V} \to \mathcal{L}$ which maps $\tilde{f}$ to F (as defined in (2)) is a unitary transform. Now we describe two important and interesting unitary transforms $\mathcal{L} \to \mathcal{L}$. Let $F := \sum_\mathbf{x} a_\mathbf{x} \xi^\mathbf{x}$ be a polynomial in $\mathcal{L}$. We define the *Hadamard transform*

$$\mathcal{H}_m(F) = \sum_{\mathbf{u} \in V_m} \hat{a}_\mathbf{u} \xi^\mathbf{u},$$

where

$$\hat{a}_\mathbf{u} = \frac{1}{\sqrt{2^m}} \sum_{\mathbf{x} \in V_m} a_\mathbf{x} (-1)^{(\mathbf{x}, \mathbf{u})},$$

i.e. we evaluate the polynomial F (now considered as a mapping) at the vector $(\xi_1, \ldots, \xi_m)$ with $\xi_i = (-1)^{u_i}$, and divide by $2^{m/2}$. We will also denote $\hat{a}_\mathbf{u}$ by $\mathcal{H}_m(F)(\mathbf{u})$. By $(\cdot, \cdot)$, we denote the standard inner product on V_m.

It is well known and easy to see that the transform $\mathcal{H}_m$ is unitary, and it can be described (after fixing an appropriate basis of $\mathcal{L}$) by the following matrix:

$$\frac{1}{\sqrt{2^m}} (\mathbf{H} \otimes \cdots \otimes \mathbf{H})$$

where

$$\mathbf{H} = \begin{pmatrix} 1 & 1 \\ 1 & -1 \end{pmatrix}.$$

We call this tensor product $\mathbf{H}_m$. If F is the indicator function of a Boolean function $f : V_m \to \mathbb{F}_2$, then

$$\mathcal{H}_m(F)(\mathbf{u}) = \frac{1}{\sqrt{2^m}} \sum_{\mathbf{x} \in V_m} (-1)^{f(\mathbf{x}) + (\mathbf{u}, \mathbf{x})}.$$

This is the classical *Hadamard-Walsh transform* of f. The function f is called *bent* if $|\mathcal{H}_m(F)(\mathbf{u})| = 1$ for all $\mathbf{u} \in V_m$. Since $\sum_\mathbf{x} (-1)^{f(\mathbf{x}) + (\mathbf{u}, \mathbf{x})} \in \mathbb{Z}$, bent functions may exist only if $\sqrt{2^m}$ is an integer, hence if m is even. Actually, for all even m bent functions do exist, see [5], for instance. The article [5] includes an excellent survey on bent functions. Another good source for classical material on bent functions is [6] or [7], for instance.

The transform $\mathcal{H}_m$ is an involution, hence we have the following well known result:

Theorem 1. *If $f : V_m \to \mathbb{F}_2$ is a bent function, then $\mathcal{H}_m((-1)^f)$ is a Boolean function, which is again bent. We call this the dual of f, denoted by $f^\perp$.*

Example 1. The Boolean function $f : V_4 \to \mathbb{F}_2$ defined by $f(\mathbf{x}) = x_1 x_2 + x_3 x_4$ is bent: The indicator polynomial of $(-1)^f$ is

$$F = 1 + \xi_1 \xi_2 \xi_3 \xi_4 + \xi_1 \xi_3 + \xi_1 \xi_4 + \xi_2 \xi_3 + \xi_2 \xi_4 + \xi_1 + \xi_2 + \xi_3 + \xi_4$$
$$- (\xi_1 \xi_2 + \xi_3 \xi_4 + \xi_1 \xi_2 \xi_3 + \xi_1 \xi_2 \xi_4 + \xi_1 \xi_3 \xi_4 + \xi_2 \xi_3 \xi_4).$$

Using just the definition of $\mathcal{H}$, we obtain

$$\mathcal{H}_4(F) = 1 + \xi_1 \xi_2 \xi_3 \xi_4 + \xi_1 \xi_3 + \xi_1 \xi_4 + \xi_2 \xi_3 + \xi_2 \xi_4 + \xi_1 + \xi_2 + \xi_3 + \xi_4$$
$$- (\xi_1 \xi_2 + \xi_3 \xi_4 + \xi_1 \xi_2 \xi_3 + \xi_1 \xi_2 \xi_4 + \xi_1 \xi_3 \xi_4 + \xi_2 \xi_3 \xi_4).$$

This is the indicator function of the dual of f. In general, it is less straightforward to compute the function $f^\perp$ that corresponds to this indicator, but it turns out that

$$f^\perp = x_1 x_2 + x_3 x_4,$$

so, in this case, f is self-dual with respect to $\mathcal{H}$.

Let $I = \sqrt{-1}$ be the complex unit. Another unitary transform $\mathcal{N}_m$ is obtained if we evaluate F at all $\pm I$-vectors $(I \cdot (-1)^{u_1}, \ldots, I \cdot (-1)^{u_m})$ of length m. We define

$$\mathcal{N}_m(F) = \sum_{\mathbf{u} \in V_m} \tilde{a}_{\mathbf{u}} \xi^{\mathbf{u}},$$

where

$$\tilde{a}_{\mathbf{u}} = \frac{1}{\sqrt{2^m}} \sum_{\mathbf{x} \in V_m} a_{\mathbf{x}} \prod_{i : x_i = 1} I^{2u_i + 1},$$

where we compute $2u_i + 1$ modulo 4.

Again, we write $\mathcal{N}_m(F)(\mathbf{u})$ instead of $\tilde{a}_{\mathbf{u}}$.

We call this transform the *nega-Hadamard transform* $\mathcal{N}_m$. In matrix terms, it is described by the m-fold tensor product

$$\frac{1}{\sqrt{2^m}} (\mathbf{N} \otimes \cdots \otimes \mathbf{N})$$

where

$$\mathbf{N} = \begin{pmatrix} 1 & I \\ 1 & -I \end{pmatrix}.$$

Another way to compute $\mathcal{N}_m(F)(\mathbf{u})$ is

$$\mathcal{N}_m(F)(\mathbf{u}) = \frac{1}{\sqrt{2^m}} \sum_{\mathbf{x} \in V_m} a_{\mathbf{x}} \cdot (-1)^{(\mathbf{u}, \mathbf{x})} \cdot I^{\text{weight}(\mathbf{x})}, \tag{3}$$

where weight($\mathbf{x}$) is the number of nonzero x_i in $\mathbf{x}$. If F is the indicator function of $(-1)^f$, this becomes

$$\mathcal{N}_m(F)(\mathbf{u}) = \frac{1}{\sqrt{2^m}} \sum_{\mathbf{x} \in V_m} (-1)^{f(\mathbf{x})+(\mathbf{u},\mathbf{x})} \cdot I^{\text{weight}(\mathbf{x})}. \tag{4}$$

A Boolean function f is called *negabent* if $|\mathcal{N}_m(F)(\mathbf{u})| = 1$ for all $\mathbf{u} \in V_m$. In contrast to bent functions, negabent functions also exist if m is odd, see Proposition 1, for instance. The difference to the case of bent functions is that there are elements $1 \pm I$ of absolute value $\sqrt{2}$ in $\mathbb{Z}[I]$, which is impossible in $\mathbb{Z}$.

The set of values $\mathcal{H}_m(F)(\mathbf{u})$ (resp. $\mathcal{N}_m(F)(\mathbf{u})$) is called the *spectrum* of F with respect to $\mathcal{H}_m$ (resp. $\mathcal{N}_m$).

Example 2. The function $f(\mathbf{x}) = x_1 x_2 + x_2 x_3 + x_3 x_4$ is bent and negabent, see Theorem 4.

Like $\mathcal{H}$, the nega-Hadamard transform is unitary: Since the polynomials $\xi^{\mathbf{x}}$, $\mathbf{x} \in V_m$, form an orthonormal basis of $\mathcal{L}$, it is sufficient to show that the polynomials $\mathcal{N}_m(\xi^{\mathbf{x}})$ are orthonormal in $\mathcal{L}$:

$$|(\mathcal{N}_m(\xi^{\mathbf{x}}), \mathcal{N}_m(\xi^{\mathbf{y}}))| = \frac{1}{2^m} \left| \sum_{\mathbf{u} \in V_m} (-1)^{(\mathbf{u},\mathbf{x})} I^{\text{weight}(\mathbf{x})} \cdot (-1)^{(\mathbf{u},\mathbf{y})}(-I)^{\text{weight}(\mathbf{y})} \right|$$

$$= \left| \frac{1}{2^m} \sum_{\mathbf{u} \in V_m} (-1)^{(\mathbf{u},\mathbf{x}+\mathbf{y})} I^{\text{weight}(\mathbf{x})-\text{weight}(\mathbf{y})} \right|$$

$$= \begin{cases} 1 \text{ if } \mathbf{x} = \mathbf{y} \\ 0 \text{ otherwise.} \end{cases}$$

Surprisingly, affine functions are negabent:

Proposition 1. *All affine functions $f : V_m \to \mathbb{F}_2$ are negabent.*

Proof. If $f(\mathbf{x}) = (\mathbf{a}, \mathbf{x})$ is linear, then the nega-Hadamard transform of the indicator of $(-1)^f$ is

$$\mathcal{N}_m((-1)^f)(\mathbf{u}) = \frac{1}{\sqrt{2^m}} \sum_{\mathbf{x} \in V_m} (-1)^{(\mathbf{u}+\mathbf{a},\mathbf{x})} \cdot I^{\text{weight}(\mathbf{x})}.$$

We define

$$\alpha := \frac{1}{\sqrt{2^{m-1}}} \sum_{\mathbf{x} \in V_m : x_1 = 0} (-1)^{(\mathbf{u}+\mathbf{a},\mathbf{x})} \cdot I^{\text{weight}(\mathbf{x})}.$$

This is $\mathcal{N}_{m-1}((-1)^g)(\mathbf{u}')$, $\mathbf{u}' = (u_2, \ldots, u_m)$, for the linear function g on V_{m-1} which is the restriction of f to $\{\mathbf{x} \in V_m : x_1 = 0\}$. By induction, we may assume $|\alpha| = 1$. Depending on $u_1 + a_1$, we get

$$\mathcal{N}_m((-1)^f)(\mathbf{u}) = \frac{1}{\sqrt{2}}(\alpha + \alpha \cdot I) \quad \text{or} \quad \frac{1}{\sqrt{2}}(\alpha - \alpha \cdot I).$$

Both numbers have absolute value 1. The induction follows by verifying that $\mathcal{N}_1((-1)^{g'})(u_m)$, for the linear function g' on V_1, has magnitude 1, both for $g' = 0$ and $g' = x_m$. Since spectral magnitudes are invariant under addition of the constant function $f(\mathbf{x}) = 1$, then affine linear functions are negabent, too. □

The next proposition will also be of interest:

Proposition 2. $\mathbf{N}_m \mathbf{H}_m \mathbf{N}_m^{-1} = (2\sqrt{2})^m \mathbf{B}_m$, where

$$\mathbf{B}_m = \begin{pmatrix} 0 & \omega \\ \overline{\omega} & 0 \end{pmatrix} \otimes \cdots otimes \begin{pmatrix} 0 & \omega \\ \overline{\omega} & 0 \end{pmatrix},$$

and $\omega = \frac{1}{\sqrt{2}}(1 + I)$ is a primitive 8-th root of unity.

Proof. Note

$$\begin{pmatrix} 1 & I \\ 1 & -I \end{pmatrix} \cdot \begin{pmatrix} 1 & 1 \\ 1 & -1 \end{pmatrix} \cdot \begin{pmatrix} 1 & 1 \\ -I & I \end{pmatrix} = \begin{pmatrix} 0 & 2(1+I) \\ 2(1-I) & 0 \end{pmatrix} = 2\sqrt{2} \cdot \begin{pmatrix} 0 & \omega \\ \overline{\omega} & 0 \end{pmatrix}$$

and "tensoring". □

In this paper, we address the following problem:

Problem 1. Find Boolean functions f that are both bent and negabent.

The main results about these objects are the following:

- For all even m, there are examples of quadratic bent-negabent functions.
- Adding a certain polynomial, c, to a bent function gives a negabent function. Adding the same polynomial, c, to a negabent function if m is even gives a bent function.
- The dual of a bent-negabent function is again bent-negabent.
- We can characterize all Maiorana-McFarland bent functions which are bent-negabent.
- We give examples of bent-negabent functions which are not quadratic.

At the end of this section, we would like to explain the connection between the transforms $\mathcal{N}_m$ and $\mathcal{H}_m$ of F and correlation properties of f, where F is the indicator polynomial of $f : V_m \to \mathbb{F}_2$ in $\mathbb{C}[\xi_1, \ldots, \xi_m]$. Note that the polynomial ring $\mathbb{C}[\xi_1, \ldots, \xi_m]$ is an algebra by the usual multiplication $\cdot$ of polynomials, which is equivalent to a convolution $*$ of the coefficients of the polynomials.

If $\mathbf{x} \in \mathbb{C}^m$, then obviously $F(\mathbf{x}) \cdot F(\mathbf{x}) = (F * F)(\mathbf{x})$. Both the Hadamard and nega-Hadamard transforms of F are nothing else than "evaluating F at certain vectors". Therefore, knowing the (nega-)Hadamard transform of F should give some information about $F * F$. We do not get full information about $F * F$, but only modulo some ideals, as we will explain now: Let I_- be the ideal in $\mathbb{C}[\xi_1, \ldots, \xi_n]$ generated by $\xi_1^2 - 1, \ldots, \xi_m^2 - 1$, and I_+ be the ideal generated by $\xi_1^2 + 1, \ldots, \xi_m^2 + 1$. Let H (resp. N) be the unique polynomial in $\mathcal{L}$ with

$H \equiv (F * F) \bmod I_-$ (resp. $N \equiv (F * F') \bmod I_+$, where F' is the polynomial in $\mathcal{L}$ whose nega-Hadamard transform is the complex-conjugate of the nega-Hadamard transform of F).

Let $\mathbf{y} \in V_m$. The coefficient $c_{\mathbf{y}}$ of $\xi^{\mathbf{y}}$ in H is the *periodic autocorrelation coefficient*

$$c_{\mathbf{y}} = \sum_{\mathbf{x} \in V_m} (-1)^{f(\mathbf{x}) + f(\mathbf{x} + \mathbf{y})}.$$

If f is bent, then $H(\mathbf{x}) = (F * F)(\mathbf{x}) = F(\mathbf{x}) \cdot F(\mathbf{x}) = 2^m$ for all $\mathbf{x} \in V_m$. Therefore, H is a polynomial such that all values in its spectrum are $\sqrt{2^m}$ (note the normalization factor $\frac{1}{\sqrt{2^m}}$). The only such polynomial is $2^m \xi^{\mathbf{0}}$, hence $c_{\mathbf{y}} = 0$ if $\mathbf{y} \neq \mathbf{0}$, and $c_{\mathbf{0}} = 2^m$, where $\mathbf{0} = (0, 0, \dots, 0)$.

Similarly, the coefficients $n_{\mathbf{y}}$ of $F * F'$ are all 0 (if $\mathbf{y} \neq \mathbf{0}$) provided f is negabent. They are called the *negaperiodic autocorrelation coefficients* of f. If $F = \sum_{\mathbf{x} \in V_m} a_{\mathbf{x}} \xi^{\mathbf{x}}$ then $F' = \sum_{\mathbf{x} \in V_m} a_{\mathbf{x}} (-1)^{\text{weight}(\mathbf{x})} \xi^{\mathbf{x}}$. Therefore, one may compute these negaperiodic autocorrelation coefficients as follows:

$$n_{\mathbf{y}} = \sum_{\mathbf{x} \in V_m} (-1)^{f(\mathbf{x}) + f(\mathbf{x} + \mathbf{y})} \cdot (-1)^{\text{weight}(\mathbf{x} + \mathbf{y})} \cdot (-1)^{(\mathbf{x}, \mathbf{y})}.$$

We need the term $(-1)^{(\mathbf{x}, \mathbf{y})}$ since our computations are modulo I_+: The inner product $(\mathbf{x}, \mathbf{y})$ counts the number of $i \in \{1, \dots, m\}$ with $x_i = y_i = 1$, which is the number of "reductions" modulo $\xi_i^2 + 1$. Every such reduction yields a "-1" since $\xi_i^2 = -1$.

The coefficient of $\mathbf{0}$ in $F * F'$ (resp. $F * F$) is called the *trivial autocorrelation coefficient*.

The following Theorem summarizes this discussion:

Theorem 2. *A Boolean function is negabent if and only if all its nontrivial negaperiodic autocorrelation coefficients are 0. It is bent, if and only if all the nontrivial periodic autocorrelation coefficients are 0.*

3 Quadratic Bent-Negabent Functions

We begin our investigation by determining quadratic bent-negabent functions. Let $\mathbf{M} = (a_{i,j})_{i,j=1,\dots,m}$ be a symmetric matrix in $\mathbb{F}_2^{(m,m)}$ with zero diagonal. Then $\mathbf{M}$ defines a quadratic function

$$p(x_1, \dots, x_n) = \sum_{i<j} a_{i,j} x_i x_j. \tag{5}$$

Conversely, any quadratic function (5) defines a symmetric matrix $\mathbf{M}$. Note that $\mathbf{M}$ may be viewed as the adjacency matrix of a graph with m vertices. If this graph is a path graph or complete (clique) graph, then we also call the corresponding quadratic function p a "path" or a "clique" function.

The following result is well known:

Theorem 3. *A quadratic function p is bent if and only if the corresponding matrix* **M** *has full rank.*

Similarly, one can characterize quadratic negabent functions. Actually, [2] contains a much more general result.

Theorem 4 ([2]). *A quadratic function p is negabent if and only if the matrix* **M** $+$ **I** *has full rank, where* **I** *is the identity matrix and* **M** *is the matrix corresponding to p.*

This theorem is the main ingredient to construct quadratic bent-negabent functions. Using a recursive formula for the determinants of matrices of the type

$$
\mathbf{L}(v_1,\ldots,v_m) := \begin{pmatrix} v_1 & 1 & & & & \\ 1 & v_2 & 1 & & & \\ & 1 & v_3 & 1 & & \\ & & \ddots & \ddots & \ddots & \\ & & & 1 & v_{m-1} & 1 \\ & & & & 1 & v_m \end{pmatrix} \in \mathbb{F}_2^{(m,m)} \tag{6}
$$

("empty" entries are 0) contained in [2], it can be shown that

$$
\det(\mathbf{L}(1,\ldots,1)) = 1
$$

if and only if $m \not\equiv 2 \bmod 3$, and

$$
\det(\mathbf{L}(0,\ldots,0)) = 1
$$

if and only if m is even. Hence the quadratic function

$$
p(x_1,\ldots,x_m) = x_1 x_2 + x_2 x_3 + \ldots + x_{m-1} x_m \tag{7}
$$

is a bent-negabent pair if m is even and $m \not\equiv 2 \bmod 6$. This result was also conjectured in [1] and proved in [3]. Theorem 5 shows that the case m even, $m \not\equiv 2 \bmod 6$ is not really exceptional. For proof, we need the following recursive construction:

Lemma 1. *Let* **A** *be a symmetric matrix in* $\mathbb{F}_2^{(m,m)}$ *such that* **A** *and* **A** $+$ **I** *have rank m. Then the matrix*

$$
\mathbf{A}' = \left(\begin{array}{c|c} \mathbf{A} & \begin{matrix} \\ 1 \end{matrix} \\ \hline \begin{matrix} & 1 \end{matrix} & \mathbf{B} \end{array} \right) \in \mathbb{F}_2^{(m+6,m+6)}
$$

with

$$
\mathbf{B} = \mathbf{L}(0,0,0,0,0,0) \in \mathbb{F}_2^{(6,6)}
$$

(the matrix of the path graph, see (6)) has rank $m+6$, *and* **A**$'$ $+$ **I** *has also rank* $m+6$.

Proof. Just do Gaussian elimination.   □

Theorem 5. *For all even $m > 2$, there exists a quadratic bent-negabent function $f : V_m \to \mathbb{F}_2$.*

Proof. For $m = 4, 6$ and 8, we take the quadratic functions corresponding to the matrices

$$
\begin{pmatrix} 0 & 1 & 0 & 0 \\ 1 & 0 & 1 & 0 \\ 0 & 1 & 0 & 1 \\ 0 & 0 & 1 & 0 \end{pmatrix},
\qquad
\begin{pmatrix} 0 & 1 & 0 & 0 & 0 & 0 \\ 1 & 0 & 1 & 0 & 0 & 0 \\ 0 & 1 & 0 & 1 & 0 & 0 \\ 0 & 0 & 1 & 0 & 1 & 0 \\ 0 & 0 & 0 & 1 & 0 & 1 \\ 0 & 0 & 0 & 0 & 1 & 0 \end{pmatrix}
$$

and

$$
\begin{pmatrix}
0 & 1 & 0 & 1 & 0 & 0 & 0 & 1 \\
1 & 0 & 1 & 0 & 0 & 0 & 0 & 0 \\
0 & 1 & 0 & 1 & 0 & 0 & 0 & 0 \\
1 & 0 & 1 & 0 & 1 & 0 & 0 & 0 \\
0 & 0 & 0 & 1 & 0 & 1 & 0 & 0 \\
0 & 0 & 0 & 0 & 1 & 0 & 1 & 0 \\
0 & 0 & 0 & 0 & 0 & 1 & 0 & 1 \\
1 & 0 & 0 & 0 & 0 & 0 & 1 & 0
\end{pmatrix}
$$

These three quadratic forms are bent-negabent since the matrices as well as the matrices plus $\mathbf{I}$ have full rank, see Theorems 3 and 4. Lemma 1 finishes the proof.   □

We will show that "adding a clique" $c(\mathbf{x}) = \sum_{i<j} x_i x_j$ (see Theorem 12) turns a bent function into a negabent function and vice versa. For quadratic functions, this has the following interpretation:

Theorem 6. *Let $\mathbf{M}$ be a symmetric matrix in $\mathbb{F}_2^{(m,m)}$, where the diagonal of $\mathbf{M}$ is zero, and let $\mathbf{J}$ denote the matrix all of whose entries are 1. Then the corresponding quadratic function is bent-negabent if and only if $\mathbf{M}$ and $\mathbf{M} + \mathbf{I} + \mathbf{J}$ have full rank.*

Proof. If f is bent, then $\mathrm{rank}(\mathbf{M}) = m$. If f is negabent, then $f + c$ is bent. The symmetric matrix that describes $f + c$ is $\mathbf{M} + \mathbf{I} + \mathbf{J}$, which must have full rank.   □

This shows that the classification of all quadratic bent-negabent functions is equivalent to the determination of all simple graphs on m vertices such that the adjacency matrix of the graph and its complement both have $\mathbb{F}_2$-rank m.

Problem 2. Determine the number of quadratic bent-negabent functions with m variables.

The following Theorem may also serve as a basic ingredient to construct many bent-negabent functions. Here $\mathbf{0}$ denotes the 0-matrix.

Theorem 7. *Let* $\mathbf{M} \in \mathbb{F}_2^{(n,n)}$ *be a symmetric matrix such that* $\mathrm{rank}(\mathbf{M}) = \mathrm{rank}(\mathbf{M} + \mathbf{J}) = n$. *Then both*

$$\mathbf{M}' = \left(\begin{array}{c|c} \mathbf{0} & \mathbf{M} \\ \hline \mathbf{M} & \mathbf{J} + \mathbf{I} \end{array} \right)$$

and $\mathbf{M}' + \mathbf{I}$ *have rank* $2n$. *Therefore, the quadratic function* f *corresponding to* $\mathbf{M}'$ *is bent-negabent.*

Proof. The matrix $\mathbf{M}'$ has maximum rank since $\mathbf{M}$ has maximum rank. Therefore, f is bent, see Theorem 3. If we add the $(2n \times 2n)$ matrix

$$\left(\begin{array}{c|c} \mathbf{I} + \mathbf{J} & \mathbf{J} \\ \hline \mathbf{J} & \mathbf{I} + \mathbf{J} \end{array} \right) .$$

to $\mathbf{M}'$, we obtain

$$\mathbf{M}'' = \left(\begin{array}{c|c} \mathbf{I} + \mathbf{J} & \mathbf{M} + \mathbf{J} \\ \hline \mathbf{M} + \mathbf{J} & \mathbf{0} \end{array} \right) .$$

Since we assume $\mathrm{rank}(\mathbf{M}) = \mathrm{rank}(\mathbf{M}+\mathbf{J}) = n$, the rank of $\mathbf{M}''$ is $2n$, which shows that f is also negabent (see Theorem 6). $\qquad\square$

An example for theorems 6 and 7 can be tested by taking $\mathbf{M} = \left(\begin{smallmatrix} 0100 \\ 1010 \\ 0101 \\ 0010 \end{smallmatrix} \right)$.

The following Theorem gives a huge family of matrices $\mathbf{M}$ of rank n such that $\mathbf{M} + \mathbf{J}$ has rank n, too. Unfortunately, n is even in this Theorem, hence we can only construct bent-negabent functions in V_m with $m \equiv 0 \bmod 4$:

Theorem 8. *Let* n *be even, and let* $\mathbf{M} \in \mathbb{F}_2^{(n,n)}$ *be a matrix where all rows and columns have odd weight (in other words,* $\mathbf{MJ} = \mathbf{JM} = \mathbf{J}$*). If* $\mathbf{M}$ *has maximum rank, then* $\mathbf{M} + \mathbf{J}$ *also has maximum rank.*

Proof. Observe that $(\mathbf{M} + \mathbf{J})^2 = \mathbf{M}^2$ and, more generally, $(\mathbf{M} + \mathbf{J})^{2s} = \mathbf{M}^{2s}$, $(\mathbf{M}+\mathbf{J})^{2s+1} = \mathbf{M}^{2s+1}+\mathbf{J}$. If $\mathbf{M}$ has even multiplicative order, then there exists s such that $\mathbf{M}^{2s} = (\mathbf{M}+\mathbf{J})^{2s} = \mathbf{I}$. Therefore, in this case $\mathbf{M}+\mathbf{J}$ has maximum rank. If $\mathbf{M}$ has odd multiplicative order, then there exists s such that $\mathbf{M}^{2s+1} = \mathbf{I}$, therefore $(\mathbf{M}+\mathbf{J})^{2s+1} = \mathbf{I}+\mathbf{J}$. But $(\mathbf{I}+\mathbf{J})^2 = \mathbf{I}$, since n is even. So, in this case, $\mathbf{M}+\mathbf{J}$ has maximum rank, too. $\qquad\square$

4   Maiorana-McFarland Bent-Negabent Functions

In this section, we briefly recall the Maiorana-McFarland construction of bent functions, and we characterize those functions which are both bent and negabent.

Let $\pi : V_n \to V_n$ be a permutation, and let $g : V_n \to \mathbb{F}_2$ be an arbitrary Boolean function. Then the function

$$\begin{aligned} f_{\pi,g} : \quad & V_m \;\to\; \mathbb{F}_2 \\ & [\mathbf{x}, \mathbf{y}] \;\mapsto\; (\mathbf{x}, \pi(\mathbf{y})) + g(\mathbf{y}) \, \text{'} \end{aligned}$$

where $m = 2n$, is bent. Here $[\cdot, \cdot]$ denotes the concatenation of vectors, and $(\cdot, \cdot)$ is the standard inner product.

We are free to choose g. If we take $g(\mathbf{y}) = y_1 \cdots y_n$, bent functions of degree $m/2$ (written as a multivariate polynomial) do exist. It is well known that this is the maximum degree. There exist bent-negabent functions of degree $m/2$ when $m = 6$ (for instance, $f(\mathbf{x}, \mathbf{y}) = x_1 y_1 y_2 + x_1 y_2 y_3 + x_2 y_1 y_2 + x_2 y_2 y_3 + x_1 y_1 + x_1 y_2 + x_2 y_3 + x_3 y_1 + x_3 y_3$ is bent-negabent), but we do not know whether bent-negabent functions of degree $m/2$ exist for all even m.

Problem 3. Find the maximum degree of bent-negabent functions.

All quadratic bent functions can be transformed by linear transformations to Maiorana-McFarland bent functions. This is a simple consequence of the fact that any quadratic function on V_m, $m = 2n$, of full rank can be transformed into $x_1' x_{n+1}' + x_2' x_{n+2}' + \ldots + x_n' x_{2n}'$ by a linear transformation

$$(x_1, \ldots, x_{2n}) \to \mathcal{L}(x_1, \ldots, x_{2n}) = (x_1', \ldots, x_{2n}').$$

However, such linear transformations do not preserve the bent-negabent property. For instance, $x_1 x_2 + x_3 x_4$ is bent, but not negabent, however $x_1 x_2 + x_2 x_3 + x_3 x_4$ is bent-negabent, see Theorem 7. These two quadratic functions are equivalent via a linear coordinate transformation. Therefore, we cannot say that all bent-negabent quadratic functions are "equivalent" to Maiorana-McFarland bent functions.

The next theorem gives a characterization of Maiorana-McFarland bent-negabent functions $f_{\pi,g}$ in terms of the permutation π and the function g:

Theorem 9. *Let $\{\mathbf{y}_1, \ldots, \mathbf{y}_{2^n}\} = V_n$, where the vectors are numbered such that*

$$\mathbf{H}_n = ((-1)^{(\mathbf{y}_i, \mathbf{y}_j)})_{i,j=1,\ldots,n}$$

is the matrix corresponding to the n-dimensional Hadamard-Walsh transform. Then $f_{\pi,g}$ is bent-negabent on V_m with $m = 2n$, if and only if all the entries in the matrix

$$\mathbf{N}_n \mathbf{P} \mathbf{D} \mathbf{N}_n^t$$

have absolute value 1, where $\mathbf{D}$ and $\mathbf{P}$ are defined as follows:

- $\mathbf{D}$ *is a diagonal matrix whose (i, i)-entry is $(-1)^{g(\mathbf{y}_i)}$.*
- $\mathbf{P}$ *is a permutation matrix where the 1-entry in row i occurs in column j and $\pi(\mathbf{y}_i) = \mathbf{y}_j$.*

Proof. We have

$\mathcal{N}_m((-1)^f)[\mathbf{u}, \mathbf{v}]$

$$= \sum_{[\mathbf{x}, \mathbf{y}] \in V_{2n}} (-1)^{(\mathbf{x}, \pi(\mathbf{y})) + g(\mathbf{y})} (-1)^{([\mathbf{u}, \mathbf{v}], [\mathbf{x}, \mathbf{y}])} I^{\text{weight}([\mathbf{x}, \mathbf{y}])}$$

$$= \sum_{\mathbf{y} \in V_n} I^{\text{weight}(\mathbf{y})} (-1)^{g(\mathbf{y})} (-1)^{(\mathbf{v}, \mathbf{y})} \left(\sum_{\mathbf{x} \in V_n} (-1)^{(\mathbf{u}, \mathbf{x})} (-1)^{(\mathbf{x}, \pi(\mathbf{y}))} I^{\text{weight}(\mathbf{x})} \right).$$

This is an entry of $\mathbf{N}_n \mathbf{H}_n \mathbf{P} \mathbf{D} \mathbf{N}_n^t$. We have $\mathbf{N}_n \mathbf{H}_n = \mathbf{B}_n \mathbf{N}_n$, where $\mathbf{B}_n$ is a diagonal matrix with all diagonal entries of absolute value 1 (see Proposition 2)), the matrix $\mathbf{N}_n\mathbf{H}_n\mathbf{PDN}_n^t$ has all entries of absolute value 1 iff $\mathbf{N}_n\mathbf{PDN}_n^t$ has this property.   □

It seems difficult to apply Theorem 9 in order to construct Maiorana-McFarland bent-negabent functions.

We say that a quadratic function p is Maiorana-McFarland if we can split the coordinates into two sets $x_1, \ldots, x_{m/2}$ and $x_{1+m/2}, \ldots, x_m$, say, such that no term $x_i x_j$ with $i \leq m/2$ and $j > m/2$ is contained in p. The following construction shows that quadratic bent-negabent functions of Maiorana-McFarland type do exist:

Theorem 10. *Let $m = 4n$, and let $\mathbf{P}$ and $\mathbf{Q}$ be permutation matrices of size n. Then the matrix*

$$\mathbf{M} = \begin{pmatrix} \mathbf{0} & \mathbf{0} & \mathbf{P} & \mathbf{0} \\ \mathbf{0} & \mathbf{0} & \mathbf{Q} & \mathbf{I} \\ \mathbf{P}^t & \mathbf{Q}^t & \mathbf{0} & \mathbf{0} \\ \mathbf{0} & \mathbf{I} & \mathbf{0} & \mathbf{0} \end{pmatrix}$$

describes a quadratic function p of Maiorana-McFarland type which is bent and negabent.

Proof. Gaussian elimination both on $\mathbf{M}$ and $\mathbf{M} + \mathbf{I}$.   □

Let $p(x_1, \ldots, x_{4n})$ be the quadratic function in Theorem 10. Numerical experiments indicate that we may always add a Boolean function $g(x_{2n+1}, \ldots, x_{3n})$ to p to obtain another bent-negabent pair. More generally still, experiments indicate that $p(x_1, \ldots, x_{4n}) = (\mathbf{y}, \phi(\mathbf{z})) + (\theta(\mathbf{z}), \mathbf{u}) + (\mathbf{u}, \mathbf{v}) + g(\mathbf{z})$, where $\mathbf{y} = (x_1, \ldots, x_n)$, $\mathbf{z} = (x_{n+1}, \ldots, x_{2n})$, $\mathbf{u} = (x_{2n+1}, \ldots, x_{3n})$, $\mathbf{v} = (x_{3n+1}, \ldots, x_{4n})$, and where $\phi, \theta : V_n \rightarrow V_n$ are both permutations, will always give examples of bent-negabent pairs on m variables from degree 2 up to degree $m/4$.

5 Transformations Which Preserve Bent-Negabentness

Theorem 11. *If f is a bent-negabent function, then its dual is again bent-negabent.*

Proof. An immediate consequence of Proposition 2.   □

There are a few more transformations which produce new bent-negabent functions from a given bent-negabent function.

Lemma 2. *Let $f : V_m \rightarrow \mathbb{F}_2$ be a bent-negabent function. Then*

1. *The Boolean function $f'(\mathbf{x}) = f(\mathbf{x}) + (\sum_{i=1}^m a_i x_i) + b$, where $(a_1, \ldots, a_m) \in V_m$, $b \in \mathbb{F}_2$, is bent-negabent.*
2. *The Boolean function $f'(\mathbf{x}) = f(x_1 + h_1, x_2 + h_2, \ldots, x_m + h_m)$ is bent-negabent.*

3. If π denotes a permutation on the set of indices $\{1, ..., m\}$, then $f(x_{\pi(1)}, ..., x_{\pi(m)})$ is bent-negabent.

Proof. We just look at (1). This is well known and also easy to see for bent functions f. The same reasoning shows that negabentness is preserved: Let $l(\mathbf{x}) = \sum_{i=1}^m a_i x_i + b$, and let $\mathbf{a} = (a_1, ..., a_m)$. Then

$$\mathcal{N}_m((-1)^{f'})(\mathbf{u}) = \sum_{\mathbf{x} \in V_m} (-1)^{f(\mathbf{x}) + (\mathbf{a}, \mathbf{x}) + b} I^{\text{weight}(\mathbf{x})} (-1)^{(\mathbf{u}, \mathbf{x})} = (-1)^b \mathcal{N}_m((-1)^f)(\mathbf{u} + \mathbf{a}).$$

$\square$

The following more interesting construction is implicitly contained in Theorem 4.6 of [4].

Theorem 12. Let $f : V_m \to \mathbb{F}_2$ be a bent function. Then $f + c$ is negabent, where $c(x_1, ..., x_n) = \sum_{i<j} x_i x_j$. Conversely, if m is even and f is negabent, then $f + c$ is bent.

Proof. Let $f : V_m \to \mathbb{F}_2$ be a bent function. Then

$$\mathcal{N}_m((-1)^{f+c})(\mathbf{u}) = \sum_{\mathbf{x} \in V_m} (-1)^{f(\mathbf{x}) + \sum_{i<j} x_i x_j} (-1)^{(\mathbf{u}, \mathbf{x})} I^{\text{weight}(\mathbf{x})}$$

$$= \sum_{\mathbf{x} \in V_m} (-1)^{f(\mathbf{x})} (-1)^{(\mathbf{u}, \mathbf{x})} I^{2c(\mathbf{x}) + \text{weight}(\mathbf{x})},$$

where we compute the exponents on the right-hand side modulo 4. We note

$$\sum_i x_i + 2 \sum_{i<j} x_i x_j = \left(\sum_i x_i\right)^2 \quad \text{in integers } \mathbb{Z}$$

where $\sum x_i = \text{weight}(\mathbf{x})$. Moreover,

$$\left(\sum_i x_i\right)^2 \equiv \begin{cases} 0 \bmod 4 & \text{if weight}(\mathbf{x}) \text{ is even} \\ 1 \bmod 4 & \text{if weight}(\mathbf{x}) \text{ is odd.} \end{cases}$$

Let E_m and O_m denote the set of vectors of even weight and odd weight, respectively. We define

$$x_e = \sum_{\mathbf{x} \in E_m} (-1)^{f(\mathbf{x})} (-1)^{(\mathbf{u}, \mathbf{x})}$$

$$x_o = \sum_{\mathbf{x} \in O_m} (-1)^{f(\mathbf{x})} (-1)^{(\mathbf{u}, \mathbf{x})}.$$

Note that both these numbers are integers. We write $\mathcal{N}_m(f + c)(\mathbf{u})$ as follows:

$$\mathcal{N}_m((-1)^{f+c})(\mathbf{u}) = x_e + I x_o$$

Now we use that f is bent, therefore

$$1 = |\mathcal{H}_m((-1)^f)(\mathbf{u})| = |x_e + x_o|$$

and

$$1 = |\mathcal{H}_m((-1)^f)(\mathbf{u} + \mathbf{j})| = |x_e - x_o|,$$

where $\mathbf{j} = (1, \ldots, 1)$ is the all-one-vector. This is only possible if $x_e = 0$ and $|x_o| = 1$, or vice versa. Therefore, $|\mathcal{N}_m((-1)^{f+c})(\mathbf{u})| = 1$.

Now let us assume that a function g is negabent. We put $f = g + c$, hence, by assumption,

$$|\mathcal{N}_m((-1)^{f+c})(\mathbf{u})| = 1 = x_e^2 + x_o^2.$$

Moreover, $x_e = n_e/\sqrt{2^m}$, and $x_o = n_o/\sqrt{2^m}$ for some integers n_e and n_o. Therefore, $n_e^2 + n_o^2 = 2^m$. If m is even, one easily shows that this is possible only if one of the two integers n_e, n_o is $\pm 2^{m/2}$, and the other is 0. Therefore, $|\mathcal{H}_m((-1)^f)(\mathbf{u})| = 1$, i.e. $f = g + c$ is bent. □

Corollary 1. *If f is a bent-negabent function, then $f + c$ is also bent-negabent.*

Remark 1. Assume that g is negabent, and m is odd. Then the proof above shows that the Hadamard-Walsh coefficients of $g + c$ are in $\{0, \pm 2^{(m+1)/2}\}$, since in this case the possible solutions of $n_e^2 + n_o^2 = 2^m$ are $n_e, n_o = \pm 2^{(m-1)/2}$. Therefore, it is possible that $n_e^2 + n_o^2 = 0$.

6   Orbits of Bent-Negabent Functions

One can view the operations that preserve the bent-negabent property, as discussed in the previous section, as generators of a group, where the action of any member of the group preserves the bent-negabent property. The two particularly interesting symmetry operations, described in Theorem 11 and Corollary 1, are involutary and, in combination with the symmetry operations of Lemma 2, generate a group of symmetries, $\mathcal{G}$, whose application on a single bent-negabent function generates an orbit of bent-negabent functions. We also consider the more trivial group, $\mathcal{E}$, generated just by the symmetries of Lemma 2. In table 1 we enumerate the number of orbits generated by the action of $\mathcal{E}$ and by $\mathcal{G}$ on the (homogeneous) quadratic Boolean functions for small numbers of variables. Note that, for quadratics, symmetry 2 of Lemma 2 is contained in symmetry 1 and does not contribute new functions to the orbit.

In a future paper we shall investigate and characterise these orbits in more detail.

Table 1. Enumeration of bent-negabent quadratic coset leaders over n variables with respect to the bent-negabent symmetry groups, $\mathcal{E}$ and $\mathcal{G}$

n	number of orbits generated by $\mathcal{E}$	number of orbits generated by $\mathcal{G}$
2	0	0
4	1	1
6	10	2
8	1272	161
10	1727780	144861

References

1. Parker, M.G.: The constabent properties of Golay-Davis-Jedwab sequences. In: IEEE Int. Symp. Inform. Theory, Sorrento, p. 302 (2000)
2. Riera, C., Parker, M.G.: Generalized bent criteria for boolean functions (I). IEEE 52, 4142–4159 (2006)
3. Riera, C., Parker, M.G.: One and Two-Variable Interlace Polynomials: A Spectral Interpretation. In: Ytrehus (ed.) WCC 2005. LNCS, vol. 3969, pp. 397–411. Springer, Heidelberg (2006)
4. Schmidt, K.U.: Quaternary constant-amplitude codes for multicode cdma. In: Submitted to IEEE Transactions on Information Theory (2006)
5. Carlet, C.: Boolean functions for cryptography and error correcting codes. In: Crama, Y., Hammer, P. (eds.) Boolean Methods and Models, Cambridge University Press, Cambridge (to appear)
6. Assmus Jr., E.F., Key, J.D.: Designs and their codes. Cambridge University Press, Cambridge (1992)
7. Golomb, S.W., Gong, G.: Signal design for good correlation. Cambridge University Press, Cambridge (2005)

Strongly Primitive Elements*

Daniel Goldstein and Alfred W. Hales

Center for Communications Research
4320 Westerra Court
San Diego, CA 92121
{dgoldste,hales}@ccrwest.org

Abstract. Let n be a positive integer. A nonzero element γ of the finite field $\mathbf{F}$ of order $q = 2^n$ is said to be "strongly primitive" if every element $(a\gamma+b)/(c\gamma+d)$, with a, b, c, d in $\{0, 1\}$ and $ad - bc$ not zero, is primitive in the usual sense. We show that the number N of such strongly primitive elements is asymptotic to $\theta\theta' \cdot q$ where θ is the product of $(1 - 1/p)$ over all primes p dividing $(q - 1)$ and θ' is the product of $(1 - 2/p)$ over the same set.

Using this result and the accompanying error estimates, with some computer assistance for small n, we deduce the existence of such strongly primitive elements for all n except $n = 1, 4,\ 6$. This extends earlier work on Golomb's conjecture concerning the simultaneous primitivity of γ and $\gamma + 1$.

We also discuss analogous questions concerning strong primitivity for other finite fields.

Keywords: finite field, primitive element, Golomb conjecture.

1 Introduction

Let $\mathbf{F} = \mathbf{F}_{2^n}$ be a finite field of order 2^n. The special linear group $SL(2, \mathbf{F}_2)$ acts on $\mathbf{F} \setminus \{0, 1\}$ by sending γ to $(a\gamma + b)/(c\gamma + d)$ for $(ad - bc) \neq 0$ with a, b, c, d in $\mathbf{F}_2 = \{0, 1\}$. We say that γ in $\mathbf{F}$ is **strongly primitive** if γ and each of its images under $SL(2, \mathbf{F}_2)$ are primitive in the usual sense, i.e. each generates the (cyclic) multiplicative group $\mathbf{F}^*$. The main object of this paper is to prove that such strongly primitive elements exist except in the cases $n = 1, 4$ and 6. This involves an asymptotic formula for the number of such strongly primitive elements, along with an error estimate, and some computer calculations to handle small cases.

In 1984 Solomon W. Golomb [4], in the process of studying Costas arrays, formulated a series of conjectures involving primitivity of elements in finite fields. These have inspired much work by many authors ([3], [6], [7], etc.). Our results can be considered as another chapter in this saga.

We also give some extensions and generalizations of our results, involving other finite fields.

---

* To Sol Golomb on the occasion of his 75th birthday.

S.W. Golomb et al. (Eds.): SSC 2007, LNCS 4893, pp. 24–36, 2007.
© Springer-Verlag Berlin Heidelberg 2007

2   First Main Result

Let $q = 2^n$. We adopt the following notation:

$W = W(q-1) = 2^k$ where k is the number of distinct prime factors p of $q-1$;

$\theta = \theta(q-1) = \phi(q-1)/(q-1) = \prod_{p|q-1}(1 - 1/p)$;

$\theta' = \theta'(q-1) = \prod_{p|q-1}(1 - 2/p)$

Note that $\theta \cdot (2^n - 1)$ is precisely the number of primitive elements in $\mathbf{F} = \mathbf{F}_{2^n}$. Also note that the image of γ under $SL(2, \mathbf{F}_2)$ consists of the six elements $\gamma, 1/\gamma, \gamma + 1, 1/(\gamma + 1), (\gamma + 1)/\gamma$, and $\gamma/(\gamma + 1)$.

Theorem 2.1. *Let N be the number of strongly primitive elements in the finite field $\mathbf{F}_{2^n}$. Then N is asymptotic to*

$$\theta\theta' \cdot 2^n$$

as $n \to \infty$.

Proof. The basic structure of this proof goes back (at least) to Carlitz [2]. First note that the function u on $\mathbf{F}_{2^n}$ which has value 1 at all primitive elements and vanishes elsewhere is given by

$$u(\gamma) = \theta \sum_{d|q-1} (\mu(d)/\phi(d)) \sum_{\mathrm{ord}(f)=d} f(\gamma) \ ,$$

where the inner sum is over all multiplicative characters f of $\mathbf{F}$ which have exact order d. Since primitivity is preserved by inversion, we need only evaluate this function u at $\gamma, \gamma + 1$ and $(\gamma + 1)/\gamma$, multiply these evaluations together, and sum over γ to evaluate N. Hence we have

$$N = \sum_\gamma u(\gamma)u(\gamma + 1)u((\gamma + 1)/\gamma)$$

$$= \theta^3 \sum_{\substack{d_1|q-1, \\ d_2|q-1, \\ d_3|q-1}} \frac{\mu(d_1)\mu(d_2)\mu(d_3)}{\phi(d_1)\phi(d_2)\phi(d_3)} \sum_{\substack{\mathrm{ord}(f)=d_1, \\ \mathrm{ord}(g)=d_2, \\ \mathrm{ord}(h)=d_3}} \sum_\gamma f(\gamma)g(\gamma + 1)h\left(\frac{\gamma + 1}{\gamma}\right).$$

We can rewrite the inner summand as $(f/h)(\gamma)(gh)(\gamma + 1)$. Hence the inner sum is equal to $K(f/h, gh)$ where $K(s,t) = \sum_\gamma s(\gamma)t(\gamma + 1)$ is the standard Jacobi sum. It is clear that $K(s,s) = q - 2$ if s is the trivial character, and that $K(s,t) = -1$ if exactly one of s,t is the trivial character. Also $K(s,t) = -1$ if st is trivial but neither s nor t is trivial. Otherwise, if none of s, t, st are trivial, it is known [5, Chapter 5, Sect. 3] that $K(s,t)$ has absolute value $\sqrt{q}$. Hence, unless s and t are both trivial, we have $|K(s,t)| \leq \sqrt{q}$.

Consider our (rewritten) expression for N. It is θ^3 times a huge sum, where each term in the sum is a coefficient times a value of $K(f/h, gh)$. First note that the sum of the absolute values of ALL the coefficients is W^3. For any particular d_1, for instance, there are $\phi(d_1)$ choices of f with exact order d_1, so the $\phi(d_1)$

in the denominator is canceled out, etc. Hence the sum in question is just the total number of triples d_1, d_2, d_3 of square-free numbers dividing $q - 1$. This is just W^3.

The main term comes from the set A of those (f, g, h) triples where $f = h = g^{-1}$. For terms in A the characters $s = f/h$ and $t = gh$ are both trivial, so the value of K is $q - 2$. For terms in A we must have $d_1 = d_2 = d_3$ and so the coefficient sum is

$$\sum_{d_1 | (q-1)} \mu(d_1)/\phi(d_1)^2 \ .$$

The sum can be rewritten as a product $\prod_{p|q-1}(1 - \frac{1}{(p-1)^2})$, which evaluates to θ'/θ^2. Therefore, the main term is equal to $\theta^3(\frac{\theta'}{\theta^2})(q - 2)$, or

$$\theta\theta' \cdot (q - 2) \ .$$

The set B is the complement of A. B consists of those triples (f, g, h) where f/h and fg are not both trivial. On B, the absolute value of K is bounded by $\sqrt{q}$. To get the error term we sum $\sum_B K$. Therefore, since the total number of terms in the sum B is $W^3 - W$, the error term E is bounded in absolute value by

$$|E| \leq \theta^3 \cdot (W^3 - W)\sqrt{q} \ .$$

Putting all this together we obtain

$$|N - \theta\theta' \cdot q| \leq (\theta W)^3 \sqrt{q} \ . \tag{2.2}$$

Since W is $O(q^\epsilon)$ for any positive ϵ, and since

$$1/2 \leq \theta'/\theta^2 \leq 1 \ , \tag{2.3}$$

we see (rearranging somewhat) that N is asymptotic to $\theta\theta' \cdot q$ as q tends to infinity. $\qquad \square$

Corollary 2.4. *Strongly primitive elements exist in $\mathbf{F}_{2^n}$ unless $n = 1, 4,$ or 6.*

Proof. We keep track of the error term E above more precisely. We have from (2.2) that $N \geq \theta\theta' \cdot q - (\theta W)^3 \sqrt{q}$, and so by (2.3) the inequality $q > 4W^6$ would be enough to guarantee that N is positive. But this always holds for $n > 228$, by Lemma 2.5, and a computer calculation shows that N is positive for all $n \leq 228$ except $n = 1, 4,$ and 6. $\qquad \square$

Lemma 2.5. *Let 2^k be the number of square-free factors of $q = 2^n - 1$. If $n > 288$ then $q > 2^{6k+2}$.*

Proof. If $k \leq 47$, then $2^{6k+2} < q$ simply because $6 \cdot 47 + 2 < 288$.

We note that (a) the product of the smallest 47 odd primes is $> 4 \cdot 64^{47}$ and that (b) the 48'th odd prime is ≥ 64. Suppose $k > 47$. Then $4 \cdot 64^k < q$ by (a) and (b). $\qquad \square$

3   Extensions

Consider now a nested pair of finite fields $\mathbf{E} \subseteq \mathbf{F}$, which henceforth are of arbitrary characteristic. Then the special linear group $SL(2, \mathbf{E})$ acts on $\mathbf{F} \setminus \mathbf{E}$ as before, but we cannot expect all images of $\gamma \in \mathbf{F}$ to be primitive. If, for instance, $|\mathbf{F}^*|$ is even (odd characteristic), then the quotient of two primitive elements cannot be primitive. So we must change the questions we ask. One possibility is to work with the full special linear group $SL(2, \mathbf{E})$ acting on $\mathbf{F} \setminus \mathbf{E}$, and modify our definitions as in Sects. 5 and 7.

Another possibility is to consider the action of a proper subgroup of $SL(2, \mathbf{E})$ such as the unipotent group of all translations $\gamma \rightarrow \gamma + c$ for $c \in \mathbf{E}$. This is (a special case of a much more general set-up) investigated by Carlitz [2] in 1956. A straightforward application of the Jacobi sum technique yields:

Theorem 3.1 (Carlitz). *Suppose* $\mathbf{E} \subseteq \mathbf{F}$ *are finite fields with* $|\mathbf{F}| = q$. *Let* $N(\mathbf{E}, \mathbf{F})$ *be the number of elements* γ *in* $\mathbf{F}$ *such that* $\gamma + c$ *is primitive for all* c *in* $\mathbf{E}$. *Then for each* $\mathbf{E}$, $N(\mathbf{E}, \mathbf{F})$ *is asymptotic to*

$$\theta^{|\mathbf{E}|} \cdot q$$

as $q \rightarrow \infty$, *where* $\theta = \theta(q - 1)$.

Hence, for large q, such γ must always exist. Here are two data points in this connection.

Proposition 3.2. *Suppose* $\mathbf{F}$ *is a quadratic extension of the finite field* $\mathbf{E}$. *Then there exists* γ *in* $\mathbf{F}$ *such that each* $\gamma + c$, c *in* $\mathbf{E}$, *is primitive, if and only if* $\mathbf{E}$ *has order 2 or 4.*

The proof of this proposition appears in Sect. 4.

Conjecture 3.3. *Let* $\mathbf{E}$ *be a finite field of characteristic* p. *Then there exists* $a \in \mathbf{E}$ *such that the polynomial* $x^p - x - a$ *is primitive (i.e. a root of* $x^p - x - a = 0$ *is primitive in the degree-p extension of* $\mathbf{E}$).

We have confirmed this by computer for all fields $\mathbf{E}$ with $|\mathbf{E}| \leq 101$. The conjecture, if true, would imply that for any field $\mathbf{F}$ of order p^p, there is a primitive $\gamma \in \mathbf{F}$ such that $\gamma + 1, \ldots, \gamma + p - 1$ are all also primitive.

4   Proof of the Proposition

In this section we prove Proposition 3.2.

Lemma 4.1. *Let* $\mathbf{C}$ *be a smooth projective genus-one curve defined over a finite field* $\mathbf{F}$. *Then the number* $|\mathbf{C}(\mathbf{F})|$ *of* $\mathbf{F}$-*rational points of* $\mathbf{C}$ *is at least* $(\sqrt{|\mathbf{F}|} - 1)^2$. *If* $|\mathbf{F}| \geq 8$, *then* $|\mathbf{C}(\mathbf{F})| > 3$.

Proof. The bound on the number of rational points is due to Siegel. The second statement follows since $(\sqrt{8} - 1)^2 > 3$. □

Proof of Proposition 3.2. Set $k = |\mathbf{E}|$. If $k = 2$, then ω and $\omega + 1$ are primitive, for ω a root of $x^2 + x + 1 = 0$. If $k = 4$, then all translates $\gamma + c, c \in \mathbf{E}$ are primitive, for γ a root of $x^2 + x + \omega = 0$. Suppose that k is odd. Then, after completing the square, we find $\beta = \gamma + c$ whose square lies in $\mathbf{E}$, for some $c \in \mathbf{E}$. Then $\beta^{2(k-1)} = 1$. As $2(k-1) < k^2 - 1$ (since $2 < k+1$), we see that β is not primitive.

Finally, suppose that $k \geq 8$ is not a power of 3. Since $3 \mid k^2 - 1$, an element of $\mathbf{F}^*$ that is equal to a cube y^3 is not primitive. There is no loss in supposing that γ does not lie in $\mathbf{E}$. Let

$$x^2 - tx + n = 0$$

be the minimal polynomial for γ over $\mathbf{E}$. Let S_γ denote the set $\{\gamma + c \mid c \in \mathbf{E}\}$ of translates of γ by elements of the smaller field $\mathbf{E}$. Let M be the number of $y \in \mathbf{F}$ whose cube y^3 lies in S_γ. Since any such y is uniquely written $u + \gamma v$ for $u, v \in \mathbf{E}$, we find, after multiplying out $(u + \gamma v)^3$ and setting the coefficient of γ equal to one, that M is the number of $u, v \in \mathbf{E}$ that satisfy:

$$1 = 3u^2 v + 3tuv^2 + (t^2 - n)v^3 \ .$$

Let $\mathbf{C} \subseteq \mathbf{P}^2$ be the homogenized curve:

$$\mathbf{C}: \qquad z^3 = 3u^2 v + 3tuv^2 + (t^2 - n)v^3 \ .$$

Then $\mathbf{C}$ is smooth of genus-1. (By a direct calculation, its discriminant is

$$\Delta = -3^9(t^2 - 4n)^2 \ ,$$

which is clearly nonzero.) By Lemma 4.1, $\mathbf{C}$ has at least four $\mathbf{E}$-rational points. Since setting $z = 0$ gives a cubic equation in u and v, there are at most 3 points on $\mathbf{C}$ "at infinity", and so $M > 0$. It follows that some $\gamma + c$ is a cube, hence not primitive, as was to be shown. $\qquad\square$

We thank Everett Howe for some assistance with genus-1 curves.

5   Relatively Strongly Primitive Elements

Definition 5.1. *An element γ in $\mathbf{F} \supseteq \mathbf{E}$ is **relatively primitive** if the (image of) γ generates the quotient group $\mathbf{F}^*/\mathbf{E}^*$. An element γ is **relatively strongly primitive (RSP)** if every image of γ under $SL(2, \mathbf{E})$ is relatively primitive.*

Clearly RSP and strongly primitive coincide if $|\mathbf{E}| = 2$.

The following generalizes our earlier theorem.

Theorem 5.2. *Suppose $\mathbf{E} \subseteq \mathbf{F}$ are finite fields, $|\mathbf{E}| = k$, $|\mathbf{F}| = q$, and the quotient $(q-1)/(k-1)$ is relatively prime to $k!$. Let $N(\mathbf{E}, \mathbf{F})$ be the number of RSP elements in $\mathbf{F}$. Then, for each $\mathbf{E}$, $N(\mathbf{E}, \mathbf{F})$ is asymptotic to*

$$(\theta_1 \theta_2 \cdots \theta_k) \cdot q$$

as $q \to \infty$, where $\theta_i = \theta_i(\frac{q-1}{k-1}) = \prod_{p \mid \frac{q-1}{k-1}} (1 - i/p)$.

Note that the condition $((q-1)/(k-1), k!) = 1$ is necessary and sufficient for no θ_i to vanish, and also necessary for any RSP elements to exist. Also note that, when $k = 2$, θ_1 coincides with our earlier θ and θ_2 coincides with θ'.

Lemma 5.3. *Let $k \geq 0$ be an integer. The number of tuples $(\gamma_0, \ldots, \gamma_k)$ in $(\mathbf{F}^*)^{k+1}$ such that each ratio γ_i/γ_j is primitive $(0 \leq i < j \leq k)$ is equal to $(q-1)^{k+1}\theta_1\theta_1 \cdots \theta_k$.*

Proof. Let ζ be a primitive element of $\mathbf{F}^*$. Then there are integers e_i such that $\gamma_i = \zeta^{e_i}$. A necessary and sufficient condition that each ratio γ_i/γ_j is primitive for $0 \leq i < j \leq k$ is that the $e_0, \ldots, e_k$ be distinct mod p for each prime divisor p of $q - 1$.

The case $k = 0$ has already been noted. Supposing the lemma true for k, we prove it for $k + 1$. Let $p \mid q - 1$, and suppose the integers $e_0, \ldots, e_k$ are distinct modulo p. Then there are $(p - k - 1)$ possibilities for e_{k+1} modulo p. Since this reasoning holds for each $p \mid q - 1$, the number of possibilities for e_{k+1} is $(q - 1) \prod_{p \mid q-1} (1 - \frac{k+1}{p})$ by the Chinese remainder theorem. $\qquad \square$

The proof of Theorem 5.2 (Sect. 11) involves standard inequalities for character sums (Sect. 10) as well as a combinatorial identity (Sect. 8) which we prove using the flow and chromatic polynomials of graph theory (Sect. 9).

6   A Second Generalization

Let $q > 1$ be a prime power. Let $\mathbf{F}$ denote a field of q elements. Let $k \geq 0$ be an integer, and let $f_0, \ldots, f_k \in \mathbf{F}[x]$ be nonzero polynomials.

We assume:

(*) The polynomials $f_0, \ldots, f_k$ are square-free and relatively prime in pairs. At most one of the f_i is constant.

For example, (*) is satisfied by the collection $\{1\} \cup \{x + c \mid c \in \mathbf{E}\}$ for $\mathbf{E}$ a subfield of $\mathbf{F}$ of cardinality k.

Note that Carlitz proved Theorem 3.1 in the more general context of a set of nonconstant polynomials satisfying (*).

We set some more notation. Let $N = N(q; f_0, \ldots, f_k)$ be the number of $\gamma \in \mathbf{F}$ such that each $f_i(\gamma)/f_j(\gamma)$ is primitive, $0 \leq i < j \leq k$. Such a γ is not a root of $h(x) = f_0(x)f_1(x) \cdots f_k(x)$. Let r (resp. R) be the number of roots of $h(x)$ in $\mathbf{F}$ (resp. in an algebraic closure of $\mathbf{F}$).

Let g be the greatest common divisor of $q - 1$ and $k!$. In Sect. 2, we defined W to be the number of square-free factors of $q - 1$. Let $\theta_i = \theta_i(q - 1)$.

If $g > 1$, then $N = 0$ and $\theta_1 \cdots \theta_k = 0$, where $\theta_i = \theta_i(q - 1)$. The first statement holds by the previous lemma. The second statement holds since the gcd condition implies that there is a prime $\ell < k$ that divides $q - 1$, hence $\theta_\ell = \prod_{p \mid (q-1)} (1 - \frac{\ell}{p}) = 0$.

Theorem 6.1. *Let $f_0(x), \ldots, f_k(x)$ be a collection of polynomials satisfying (*).*

1. *If $g > 1$ then $\theta_1 \theta_2 \cdots \theta_k = N = 0$.*
2. *$|N - (q - r)\theta_1 \cdots \theta_k| \leq (\theta W)^{k(k+1)/2} (R - 1)\sqrt{q}$.*
3. *We vary the field. Assume $g = 1$, and let $N_n = N(q^n; f_0, \cdots, f_k)$. Then the ratio*

$$\frac{N_n}{\theta_1 \cdots \theta_k q^n} \to 1$$

as $n \to \infty$ through a sequence of integers so that $(q^n - 1, k!) = 1$, and where $\theta_i = \theta_i(q^n - 1)$.

The theorem as stated is of limited applicability. Because of the gcd condition, only fields of characteristic 2 are allowed, and even then, the gcd condition is quite restrictive. However, in Sect. 7 we state a relative version of Theorem 6.1 which applies to many more fields (at the price of only having a "relative" conclusion).

We note that the assumption (*) can be replaced by much weaker conditions. For example, it is enough that there exist irreducible polynomials $\pi_i, 1 \leq i \leq k$ in $\mathbf{F}[x]$ such that $\pi_i(x) \mid f_j(x)$ for $0 \leq j \leq k$ if and only if $i = j$, and $\pi_i^2(x)$ does not divide $f_i(x)$. This condition can be weakened even further:

(**) For all primes p not dividing q, if the integers $e_0, \ldots, e_k$ satisfy $e_1 + \cdots + e_k \equiv 0 \pmod{p}, c \in \mathbf{F}^*$, and if

$$f_0(x)^{e_0} f_1(x)^{e_1} \ldots f_k(x)^{e_k} = cg(x)^p$$

is a nonzero constant multiple of a pth power of a polynomial $g(x) \in \mathbf{F}[x]$, then each $e_0, \ldots, e_k$ is divisible by p.

7   A Relative Version

We reformulate Theorem 6.1 in a "relative" way. Theorem 7.2 simultaneously generalizes Theorems 2.1, 5.2, and 6.1.

Definition 7.1. *Let μ be a subgroup of $\mathbf{F}^*$. Set $m = |\mu|$. An element γ in $\mathbf{F}$ is **relatively primitive with respect to** μ if the (image of) γ generates the quotient group $\mathbf{F}^*/\mu$.*

Let $N = N(q; f_0, \ldots, f_k; \mu)$ denote the number of $\gamma \in \mathbf{F}$ such that $f_i(\gamma) \neq 0$ for all $0 \leq i \leq k$ and $f_i(\gamma)/f_j(\gamma)$ is relatively primitive with respect to μ for all $0 \leq i < j \leq k$.

Let g be the greatest common divisor of $(q - 1)/m = |\mathbf{F}^*/\mu|$ and $k!$. Let $\theta_i = \theta_i((q - 1)/m) = \prod_{p \mid \frac{q-1}{m}} (1 - i/p)$.

Theorem 7.2. *Let $q > 1$ be a prime power, $\mathbf{F}$ a field of cardinality q. Let the polynomials $f_0, \cdots, f_k \in \mathbf{F}[x]$ satisfy (*) or (**).*

1. *If $g > 1$ then $\theta_1 \theta_2 \cdots \theta_k = N = 0$.*
2. *$|N - (q - r)\theta_1 \cdots \theta_k| \leq (\theta W)^{k(k+1)/2}(R - 1)\sqrt{q}$.*
3. *We vary the field. Let μ be a subgroup of $\mathbf{F}^*$, $N_n = N(q^n; f_0, \cdots, f_k; \mu)$. The ratio*

$$\frac{N_n}{\theta_1 \cdots \theta_k q^n} \to 1$$

as $n \to \infty$, through a sequence of values such that $((q^n - 1)/m, k!) = 1$, where now $\theta_i = \theta_i(\frac{q^n - 1}{m})$.

8 A Key Formula

We state a combinatorial formula that is the key to proving Theorem 7.2.

Let $k \geq 1$ be an integer and p a prime. We define a system of $k + 1$ linear equations in $k(k + 1)/2$ unknowns $y_{i,j}, 0 \leq i < j \leq k$ over the field $\mathbf{Z}/p\mathbf{Z}$. For each $0 \leq i_0 \leq k$, we have the equation (E_{i_0}) :

$$(E_{i_0}) : \qquad \sum_{i_0 < j \leq k} y_{i_0, j} = \sum_{0 \leq j < i_0} y_{j, i_0} .$$

Let Y be the set of simultaneous solutions to $(E_0), (E_1), \ldots, (E_k)$. For each $y = (y_{i,j}) \in Y$, let $z(y)$ denote the number of i, j pairs such that $y_{i,j} = 0$.

We have the following key formula:

$$\frac{1}{(-p)^{k(k+1)/2}} \sum_{y \in Y} (1 - p)^{z(y)} = (1 - \frac{1}{p})(1 - \frac{2}{p}) \cdots (1 - \frac{k}{p}) , \qquad (8.1)$$

whose proof will be postponed to the next section.

9 Some Graph Theory

We state and prove a result on polynomial invariants of graphs. Our sole purpose is to give a proof of (8.1). The reader who is willing to take that formula on faith (or who has an alternative proof) can skip this section.

Let $G = (E, V)$ be a graph. Choose an arbitrary orientation to each edge in E. A labeling of the edges of G with elements of the finite abelian group H is a **flow** if the **divergence** (the out-flow minus the in-flow) at each vertex is equal to 0.

The number of H-flows on G is independent of the orientation put on G, and is therefore a function of G and H alone. As Tutte showed [8, pp. 55–69], this number is the same for all abelian groups H of the same order n. Without loss of generality, we restrict our attention to $\mathbf{Z}/n\mathbf{Z}$, the additive group of integers mod n.

For $A \subseteq E$ a set of edges, let $\tilde{\mathrm{fl}}(A, n)$ denote the number of flows with values in $\mathbf{Z}/n\mathbf{Z}$ that are supported on A.

It is clear that $\tilde{\mathrm{fl}}(A, n)$ is a power of n, namely,

$$\tilde{\mathrm{fl}}(A, n) = n^{|A|-\rho(A)} , \tag{9.1}$$

where the rank of A, $\rho(A)$, is the number of edges in a maximal acyclic subset of A. For $A \subseteq E$ a subset of edges, the **flow number** $\mathrm{fl}(A, n)$ is the number of $\mathbf{Z}/n\mathbf{Z}$-flows on G with support precisely equal to A.

We change gears somewhat and write $\mathrm{fl}(A, \lambda)$, where λ is a variable, to indicate that this is a polynomial (rather than a function of a nonnegative integer n).

The flow polynomial satisfies the recurrence:

$$\mathrm{fl}(A, \lambda) = \sum_{B \subseteq A} (-1)^{|A|-|B|} \tilde{\mathrm{fl}}(B, \lambda) . \tag{9.2}$$

The chromatic number $\mathrm{chrom}(G; n)$ is the number of ways of assigning n colors to the vertices of G so that no two adjacent vertices have the same color. The chromatic polynomial satisfies the recurrence:

$$\mathrm{chrom}(G; \lambda) = \sum_{A \subseteq E} (-1)^{|A|} \lambda^{|V|-\rho(A)} . \tag{9.3}$$

Equations (9.2) and (9.3) are each proved by a straightforward inclusion-exclusion argument that we omit. They imply that the flow and chromatic polynomials really are polynomials. (Alternatively, this follows from the fact that the flow and chromatic polynomials are specializations of the two-variable Tutte polynomial, see [1, Proof of Theorem 14.1 and Exercise 14e]).

Lemma 9.4. *Let $G = (V, E)$ be a graph. Then*

$$(-1)^{|E|} \sum_{A \subseteq E} (1 - \lambda)^{|E|-|A|} \mathrm{fl}(A, \lambda) = \lambda^{|E|-|V|} \mathrm{chrom}(G; \lambda) .$$

Proof. The left hand side of the formula is equal to

$$= (-1)^{|E|} \sum_{A \subseteq E} (1 - \lambda)^{|E|-|A|} \sum_{B \subseteq A} (-1)^{|A|-|B|} \tilde{\mathrm{fl}}(B, \lambda) \quad \text{(by (9.2))}$$

$$= (-1)^{|E|} \sum_{B \subseteq E} (-\lambda)^{|E|-|B|} \tilde{\mathrm{fl}}(B, \lambda) \quad \text{(by the binomial theorem)}$$

$$= \sum_{B \subseteq E} (-1)^{|B|} \lambda^{|E|-\rho(B)} \quad \text{(by (9.1))}$$

$$= \lambda^{|E|-|V|} \mathrm{chrom}(G; \lambda).$$

$\square$

We now show how the key formula follows from the lemma. The complete graph $K_{k+1} = (V, E)$ on $k + 1$ vertices has $|E| = k(k + 1)/2$ and $|V| = k + 1$. The chromatic polynomial of K_{k+1} is easily seen to be

$$\mathrm{chrom}(G, \lambda) = \lambda(\lambda - 1) \cdots (\lambda - k) .$$

Setting $\lambda = p$, in Lemma 9.4, we have:

$$(-1/p)^{k(k+1)/2} \sum_{A \subseteq E} (1-p)^{|E|-|A|} \mathrm{fl}(A,p) = p^{-k-1} p(p-1) \cdots (p-k) .$$

Using the definition of $\mathrm{fl}(A,p)$ and rearranging somewhat yields (8.1), the exact form that will be used in Sect. 11.

10   Character Sums

In this section we state (but do not prove) the character sum bound needed for Theorem 7.2. Let $f(x)$ be a polynomial in $\mathbf{F}[x]$, ψ a multiplicative character of $\mathbf{F}^*$ taking complex values, m the order of ψ. As a particular case of a result of Weil [9], the character sum

$$\sum_{\gamma \in \mathbf{F}} \psi(f(\gamma))$$

can be nicely bounded in terms of R, the number of distinct roots of f in an algebraic closure, provided $f(x)$ is not (up to nonzero scalar multiple) an m'th power.

Lemma 10.1. *([5, Theorem 5.41]) Let $\mathbf{F}$ be a finite field of q elements, ψ a multiplicative character of $\mathbf{F}^*$ of order $m > 1$, $f \in \mathbf{F}[x]$ a nonconstant polynomial such that $f(x)/c$ is not the m'th power of a polynomial, where c is the leading coefficient of $f(x)$. Then*

$$\left| \sum_{\gamma \in \mathbf{F}} \psi(f(\gamma)) \right| \leq (R-1)\sqrt{q} .$$

Note that the inequality for Jacobi sums used above, namely, $|K(s,t)| \leq \sqrt{q}$ if s or t is nonidentity, follows from Lemma 10.1 where $f(x) = x^a(x+1)^b$, ψ primitive, and $R = 2$.

11   Proof of the Theorem

We have $q > 1$ a prime power, $\mathbf{F}$ a field of q elements, $f_0(x), \ldots, f_k(x) \in \mathbf{F}[x]$ satisfying either (*) or the weaker condition (**). Let $N = N(q; f_0, \cdots, f_k)$ be the number of $\gamma \in \mathbf{F}$ such that each ratio $f_i(\gamma)/f_j(\gamma)$ is primitive for $0 \leq i < j \leq k$.

We have

$$N = \sum_{\gamma} \prod_{0 \leq i < j \leq k} u\left(\frac{f_i(\gamma)}{f_j(\gamma)}\right) ,$$

the sum being over those $\gamma \in \mathbf{F}$ that are not roots of $f_0(x) \cdots f_k(x)$. This consideration excludes r values from the sum, where r is the number of roots in $\mathbf{F}$ of $f_0(x) \cdots f_k(x)$.

Let $d_{i,j}$ be a divisor of $q-1$, $\chi_{i,j}$ a character of $\mathbf{F}^*$ of exact order $d_{i,j}$ for all pairs $0 \le i < j \le k$. In view of the definition of u, we have

$$N = \theta^{k(k+1)/2} \sum_{\gamma} \sum_{d_{i,j}|(q-1)} \left(\left(\prod_{0 \le i < j < k} \frac{\mu(d_{i,j})}{\phi(d_{i,j})} \right) \sum_{\text{ord}\,\chi_{i,j}=d_{i,j}} \prod_{0 \le i < j < k} \chi_{i,j}\left(\frac{f_i(\gamma)}{f_j(\gamma)} \right) \right) .$$

We are free to interchange the order of summation, to get

$$N = \theta^{k(k+1)/2} \sum_{d_{i,j}|(q-1)} \left(\prod \frac{\mu(d_{i,j})}{\phi(d_{i,j})} \right) \sum_{\text{ord}\,\chi_{i,j}=d_{i,j}} \sum_{\gamma} \prod_{i<j} \chi_{i,j}\left(\frac{f_i(\gamma)}{f_j(\gamma)} \right) .$$

The inner sum (over γ) can be rewritten as

$$\sum_{\gamma} \prod_{i_0} \left(\prod_{i_0 < j \le k} \chi_{i_0,j}(f_{i_0}(\gamma)) \Big/ \prod_{0 \le j < i_0} \chi_{j,i_0}(f_{i_0}(\gamma)) \right) .$$

As before, we estimate N with a main term and an error term. The main term is summed over the set A consisting of those collections of characters $(\chi_{i,j} : 0 \le i < j \le k)$ such that, for each $0 \le i_0 \le k$, we have

$$(***) \qquad \prod_{i_0 < j \le k} \chi_{i_0,j} = \prod_{0 \le j < i_0} \chi_{j,i_0} .$$

For each collection of characters in the main term A, the innermost sum is equal to $q - r$ (compare with $q - 2$ in Sect. 2).

Hence, the main term is $(q-r)M$, where

$$M = \theta^{k(k+1)/2} \sum_{d_{i,j}|(q-1)} \left(\prod \frac{\mu(d_{i,j})}{\phi(d_{i,j})} \right) \sum_{\substack{\text{ord}\,\chi_{i,j}=d_{i,j}, \\ (***)}} 1 ,$$

where the innermost sum is over those characters of the stated order that also satisfy the equations (***).

It is not hard to see that M is a product of local factors

$$M = \prod_{p|(q-1)} M_p ,$$

where M_p is $(1 - \frac{1}{p})^{k(k+1)/2}$ times the sum above restricted to those collections of divisors $(d_{i,j} : 0 \le i < j \le k)$, with each $d_{i,j} = 1$ or p. For such a collection, we see that

$$\prod_{0 \le i < j \le k} \frac{\mu(d_{i,j})}{\phi(d_{i,j})}$$

is equal to $(1-p)^{-\,\text{nz}(d)}$, where $\text{nz}(d)$ is the number of $d_{i,j}$ that are not equal to 1. We get

$$M_p = \left(\frac{p-1}{p} \right)^{k(k+1)/2} \sum_{d_{i,j}|p} (1-p)^{-\,\text{nz}(d)} \sum_{\substack{\text{ord}\,\chi_{i,j}=d_{i,j}, \\ (***)}} 1 .$$

Combining powers of $(1 - p)$ gives:

$$M_p = \frac{1}{(-p)^{k(k+1)/2}} \sum_{d_{i,j}|p} (1-p)^{z(d)} \sum_{\substack{\text{ord}\, \chi_{i,j}=d_{i,j}, \\ (***)}} 1 \ , \tag{11.1}$$

where $z(d) = k(k+1)/2 - \text{nz}(d)$ is the number of $d_{i,j}$ that are equal to 1. Then:

$$M_p = \frac{1}{(-p)^{k(k+1)/2}} \sum_{\text{ord}\, \chi_{i,j}|p} (1-p)^{z(\chi)} \ , \tag{11.2}$$

where $z(\chi)$ is the number of $0 \le i < j \le k$ such that $\chi_{i,j} = 1$.

We identify the cyclic multiplicative group of pth roots of unity in $\mathbf{F}$ with the additive cyclic group $\mathbf{Z}/p\mathbf{Z}$ of order p. Choose a primitive p^{th} root of unity ζ and define $y_{i,j}$ by the formula

$$\chi_{i,j} = \zeta^{y_{i,j}} \ .$$

Then the $y_{i,j}$ satisfy $(E_0), \ldots, (E_k)$ since the $\chi_{i,j}$ satisfy $(***)$, so that the sum (11.2) is equal to (8.1). We thus get for the main term:

$$M_p = (1 - \frac{1}{p})(1 - \frac{2}{p}) \cdots (1 - \frac{k}{p}), \quad \text{and} \tag{11.3}$$

$$(q - r)M = (q - r)\theta_1\theta_2 \cdots \theta_k, \tag{11.4}$$

and we turn now to the error term.

The error term E is obtained by summing those terms where the character is not trivial. Let ζ be a primitive character of $\mathbf{F}^*$. Then any character $\chi_{i,j}$ is equal to $\zeta^{e_{i,j}}$ for some integer $e_{i,j}$. View the $e_{i,j}$ as a labeling on the edges of the complete graph K_{k+1}. Finally, let the integer $D_i(0 \le i < k)$ denote the divergence at vertex i corresponding to the labeling of the (directed) edges of K_{k+1} with the labels $e_{i,j}(0 \le i < j \le k)$.

We obtain:

$$E = \theta^{k(k+1)/2} \sum_{d_{i,j}} \prod_{0 \le i < j \le k} \frac{\mu(d_{i,j})}{\phi(d_{i,j})} \sum_{\substack{\text{ord}\, \chi_{i,j}=d_{i,j}, \\ \text{not}\ (***)}} \sum_{\gamma} \zeta\left(f_0(\gamma)^{D_0} \cdots f_k(\gamma)^{D_k}\right) \ .$$

We are free to modify the inner rational function by multiplying by an appropriate power of $(f_0(x) \cdots f_k(x))^{q-1}$. This does not change the value of the rational function. Call the resulting polynomial $f(x)$. By condition $(**)$ (or, by condition $(*)$, which implies $(**)$), we see that $f(x)$ is not a nonzero scalar multiple of an mth power of a polynomial, $m > 1$, so that Lemma 10.1 applies. Therefore, the inner sum is at most $(R - 1)\sqrt{q}$ in absolute value. Since the number of terms is at most $W^{k(k+1)/2}$ and each coefficient has absolute value 1, we get an error term of

$$|E| \le (\theta W)^{k(k+1)/2}(R - 1)\sqrt{q} \ . \tag{11.5}$$

Theorem 6.1 follows from (11.4) and (11.5).

The proof of the "relative" version (Theorem 7.2) is exactly the same, except that the main term is now a product over those primes p dividing $(q - 1)/m$ (rather than all primes dividing $q - 1$).

References

1. Biggs, N.: Algebraic graph theory, 2nd edn. Cambridge Mathematical Lectures. Cambridge University Press, Cambridge (1993)
2. Carlitz, L.: Sets of primitive roots. Compositio Math. 13, 65–70 (1956)
3. Cohen, S.D., Mullen, G.L.: Primitive elements in finite fields and Costas arrays. Appl. Algebra Engrg. Comm. Comput. 2(1), 45–53 (1991)
4. Golomb, S.W.: Algebraic constructions for Costas arrays. J. Combin. Theory Ser. A 37(1), 13–21 (1984)
5. Lidl, R., Niederreiter, H.: Finite fields, 2nd edn. Encyclopedia of Mathematics and its Applications, vol. 20. Cambridge University Press, Cambridge (1997)
6. Moreno, O.: On the existence of a primitive quadratic of trace 1 over $GF(p^m)$. J. Combin. Theory Ser. A 51(1), 104–110 (1989)
7. Moreno, O., Sotero, J.: Computational approach to conjecture A of Golomb. In: Proceedings of the Twentieth Southeastern Conference on Combinatorics, Graph Theory, and Computing (Boca Raton, FL, 1989), vol. 70, pp. 7–16 (1990)
8. Tutte, W.T.: Selected papers of W. T. Tutte, vol. I. McCarthy, D., Stanton, R.G. (eds.). Charles Babbage Research Center, Winnipeg (1979)
9. Weil, A.: On some exponential sums. Proc. Nat. Acad. Sci. 34, 204–207 (1948)

The Perfect Binary Sequence of Period 4 for Low Periodic and Aperiodic Autocorrelations*

Nam Yul Yu and Guang Gong

Department of Electrical and Computer Engineering
University of Waterloo, Waterloo, Ontario, Canada
yny0628@gmail.com, ggong@calliope.uwaterloo.ca

Abstract. The perfect binary sequence of period 4 − '0111' (or cyclic shifts of itself or its complement) − has the optimal periodic autocorrelation function where all out-of-phase values are zero. Not surprisingly, it is also the Barker sequence of length 4 where all out-of-phase aperiodic autocorrelation values have the magnitudes of at most one. From these observations, the applications of the sequence for low periodic and aperiodic autocorrelations are studied. First, the perfect sequence is discussed for binary sequences with optimal periodic autocorrelation. New binary sequences of period $N = 4(2^m − 1), m = 2k$ with optimal periodic autocorrelation are presented, which are obtained by a slight modification of product sequences of binary m-sequences and the perfect sequence. Then, it is observed that a product sequence of the Legendre and the perfect sequences has not only the optimal periodic but also the good aperiodic autocorrelations with the asymptotic merit factor 6. Moreover, if the product sequences replace Legendre sequences in Borwein, Choi, Jedwab (BCJ) sequences, or equivalently Kristiansen-Parker sequences (simply BCJ-KP sequences), numerical results show that the resulting sequences have the same asymptotic merit factor as the BCJ-KP sequences.

Keywords: Binary sequences, Merit factors, Perfect sequence, Periodic and aperiodic autocorrelations.

1 Introduction

In code-division multiple access (CDMA) [1] [2], wireless local area network (WLAN) [3], and ultra-wideband radio (UWB) [28], binary pseudorandom sequences with good autocorrelation functions play important roles for desired power spectrums, synchronizations, etc.

Conventional autocorrelation functions have two different definitions − periodic and aperiodic autocorrelations. Traditionally, the periodic autocorrelation of a binary sequence has received more attention in literatures for sequence families of CDMA communication systems. However, the aperiodic autocorrelation is considered to better characterize a binary sequence for more realistic communication systems. For further discussions of the autocorrelation functions and their applications, see [25].

---

* This work was supported by NSERC Grant RGPIN 227700-00.

S.W. Golomb et al. (Eds.): SSC 2007, LNCS 4893, pp. 37–49, 2007.

The *perfect* binary sequence of period 4 is a uniquely known binary sequence where all out-of-phase periodic autocorrelation values are zero. Not surprisingly, it is also the Barker sequence of length 4 where all out-of-phase aperiodic autocorrelation values have the magnitudes of at most 1. From these observations, we consider the perfect sequence is *optimal* in terms of both periodic and aperiodic autocorrelations, and believe the sequence can be effectively utilized for constructing binary sequences with low periodic or aperiodic autocorrelations.

In this paper, we study the applications of the perfect binary sequence for both definitions of autocorrelation functions. A basic approach for the applications is a product method described in [19]. First, we consider the usage of the perfect sequence in the product sequences which have optimal periodic autocorrelation. Then, we present new binary sequences with optimal periodic autocorrelation magnitudes by slightly modifying product sequences of binary m-sequences and the perfect sequence. In terms of aperiodic autocorrelation, on the other hand, numerical results show that product sequences of Legendre sequence with the perfect sequence preserve the asymptotic merit factor 6. Moreover, it is also observed if the product sequences replace Legendre or modified Jacobi sequences in Borwein, Choi, Jedwab (BCJ) sequences [8], or equivalently Kristiansen-Parker sequences [17] [18] (simply BCJ-KP sequences), the resulting sequences have the same asymptotic merit factor 6.34 as the BCJ-KP sequences, which is known to be the highest value for binary sequences generated by a constructive way [8] [17] [18].

From theoretical proofs and the numerical evidences, we conclude that the perfect binary sequence is very useful for obtaining new sequences with low periodic and aperiodic autocorrelations.

2   Binary Sequences with Low Autocorrelation

2.1   Binary Sequences with Low Periodic Autocorrelation

The periodic autocorrelation of a binary sequence $\mathbf{a} = \{a_i\}$ of period N is defined by

$$C_{\mathbf{a}}(\tau) = \sum_{i=0}^{N-1} (-1)^{a_i + a_{i+\tau}}, \quad 0 \leq \tau \leq N - 1$$

where the indices are computed modulo N. For a sequence $\mathbf{a}$ of period N, it is implied that $C_{\mathbf{a}}(\tau) = N$ occurs only at $\tau \equiv 0 \pmod{N}$. $C_{\mathbf{a}}(\tau)$ is called *optimal* [4] if it satisfies

1) $C_{\mathbf{a}}(\tau) \in \{N, -1\}$ if $N \equiv 3 \pmod{4}$, or
2) $C_{\mathbf{a}}(\tau) \in \{N, 1, -3\}$ if $N \equiv 1 \pmod{4}$, or
3) $C_{\mathbf{a}}(\tau) \in \{N, 2, -2\}$ if $N \equiv 2 \pmod{4}$, or
4) $C_{\mathbf{a}}(\tau) \in \{N, 0, -4\}$ or $\{N, 0, 4\}$ if $N \equiv 0 \pmod{4}$

for all τ's. For complete classes of known binary sequences with optimal periodic autocorrelation, see [9] and [31].

2.2   Binary Sequences with Good Aperiodic Autocorrelation

The aperiodic autocorrelation of a binary sequence $\mathbf{a} = \{a_i\}$ of length N is defined by

$$A_{\mathbf{a}}(\tau) = \sum_{i=0}^{N-\tau-1} (-1)^{a_i + a_{i+\tau}}, \quad 0 \leq \tau \leq N - 1.$$

In the design of binary sequences, the low aperiodic autocorrelation is known to be much more difficult to achieve than the low periodic autocorrelation [15]. In general, Barker sequences [5] have the best known aperiodic autocorrelation functions where all out-of-phase values have the magnitudes of at most 1. Due to the strict condition, however, there exist only a few Barker sequences of small lengths − $N = 2, 3, 4, 5, 7, 11$, and 13. Furthermore, it is shown that there are no further Barker sequences of odd lengths [29].

As a mild and alternative measure of aperiodic autocorrelation, the *merit factor* [10] of a binary sequence $\mathbf{a}$ of length N is defined by

$$F_N(\mathbf{a}) = \frac{N^2}{2 \sum_{\tau=1}^{N-1} [A_{\mathbf{a}}(\tau)]^2}.$$

The merit factor measures how collectively small the aperiodic autocorrelation is. Also, it is a measure of the spectral uniformity of the sequence [7], which is of interest in digital communications. Let $F(\mathbf{a}) = \lim_{N \to \infty} F_N(\mathbf{a})$ be the asymptotic merit factor of a binary sequence $\mathbf{a}$ as its length goes to infinity. Then, a major design issue on the merit factor is to find binary sequences of length N with high asymptotic merit factors. The best known and theoretically proven asymptotic merit factor of binary sequences generated by constructive ways is 6. For more details of the binary sequences with high merit factors, see [15], [23], and [31].

2.3   The Perfect Sequence

Let $\mathbf{a}$ be a binary sequence of period N. If its periodic autocorrelation $C_{\mathbf{a}}(\tau)$ is equal to 0 for all $\tau \not\equiv 0 \pmod{N}$, i.e.,

$$C_{\mathbf{a}}(\tau) = \begin{cases} 0, & \text{if } \tau \not\equiv 0 \bmod N \\ N, & \text{if } \tau \equiv 0 \bmod N, \end{cases}$$

then $\mathbf{a}$ is called the *perfect sequence*. The only known perfect binary sequence is $\mathbf{a} = (0, 1, 1, 1)$ or its complement [6]. For a period of $4 < N < 108900$, no perfect binary sequences are found [27], and it is conjectured in [16] that no other perfect binary sequences exist except for $N = 4$.

It is easy to see that any out-of-phase aperiodic autocorrelation value of the perfect binary sequence is $1, 0$, or -1. Therefore, the perfect sequence is a Barker sequence as well. From this point of view, we consider the sequence is *optimal* not only in periodic but also in aperiodic autocorrelations.

3 The Perfect Sequence for Binary Sequences with Optimal Periodic Autocorrelation

In this section, we discuss the applications of the perfect binary sequence for binary sequences with optimal periodic autocorrelation. First, the product [19] sequences are described as the application example. Then, binary sequences of period $N = 4(2^m - 1), m = 2k$ with optimal periodic autocorrelation magnitudes [30] are described by the slight modification of a product sequence.

3.1 Product Sequences

Let $\mathbf{a}$ and $\mathbf{b}$ be binary sequences of periods N_1 and N_2, respectively, where $\gcd(N_1, N_2) = 1$. Then the *product sequence* [19] $\mathbf{p} = \mathbf{a} + \mathbf{b} = (p_0, p_1, \cdots, p_{N-1})$ of period $N = N_1 N_2$ is defined by the component-wise addition of $p_i = a_i + b_i$ (mod 2), $0 \le i \le N - 1$. Periodic autocorrelation of the product sequence is given by

$$
\begin{aligned}
C_{\mathbf{p}}(\tau) = \sum_{i=0}^{N-1}(-1)^{p_{i+\tau}+p_i} &= \left[\sum_{i_1=0}^{N_1-1}(-1)^{a_{i_1+\tau}+a_{i_1}}\right] \cdot \left[\sum_{i_2=0}^{N_2-1}(-1)^{b_{i_2+\tau}+b_{i_2}}\right] \\
&= C_{\mathbf{a}}(\tau) \cdot C_{\mathbf{b}}(\tau), \quad 0 \le \tau \le N - 1
\end{aligned}
\tag{1}
$$

where the indices of a sequence are computed modulo its own period. From (1), we could obtain binary product sequences with optimal periodic autocorrelation using the perfect sequence.

Application 1. *[19] Let $\mathbf{a} = (0, 1, 1, 1)$ be the perfect binary sequence of period 4. Let $\mathbf{b}$ be a binary sequence of period v with ideal two-level autocorrelation, where $\gcd(4, v) = 1$. In other words, $\mathbf{b}$ is*

1) *a binary sequence of period $v = 2^m - 1$ with ideal two-level autocorrelation listed in [9], or*
2) *a Legendre sequence [26] of period $v \equiv 3$ (mod 4) where v is prime, or*
3) *a twin-prime sequence [16] of period $v = pq$ where p and $q = p + 2$ are distinct primes, or*
4) *a Hall's sextic residue sequence [13] of period $v = 4x^2 + 27$ where v is prime.*

Then, a binary product sequence $\mathbf{p} = \mathbf{a} + \mathbf{b}$ of period $4v$ has optimal periodic autocorrelation from (1), i.e., $C_{\mathbf{p}}(\tau) \in \{N, 0, -4\}$.

3.2 Binary Sequences with Optimal Periodic Autocorrelation Magnitude

In [30], Yu and Gong presented binary sequences of period $N = 4(2^m - 1)$, $m = 2k$, $k > 1$ with optimal four-valued periodic autocorrelation by slightly modifying the product sequences in Application 1 of binary m-sequences and the perfect sequence. Before introducing the sequences, we first define the concept of the constant-on-cosets property [12] of binary sequences.

Definition 1. *Let* $\mathbf{a} = \{a_i\}$ *be a binary sequence of period* N. *Then,* $\mathbf{a}$ *is said to have the* constant-on-cosets *property if* $a_i = a_{2i}$ *for all* i'*s, where the index is computed modulo* N.

Application 2. *[30] Let* $\mathbf{a} = (0, 1, 1, 1)$ *be the perfect binary sequence of period 4 and* $\mathbf{b} = \{b_i\}$ *a binary* m-*sequence of period* $2^m - 1$, $m = 2k$, $k > 1$ *with the* constant-on-cosets *property. Also, let* $\mathbf{c} = \{c_i\}$ *be a binary sequence of period* $n = 4(2^k + 1)$ *defined by*

$$c_i = \begin{cases} 0, & \text{if } i \neq i'(2^k + 1), \\ z_{i'}, & \text{if } i = i'(2^k + 1) \end{cases}$$

where i' *is an integer,* $0 \leq i' \leq 3$ *and* $\mathbf{z} = (z_0, z_1, z_2, z_3)$ *is any cyclic shift of* $(1, 1, 0, 0)$. *Let* $\mathbf{u}$ *be a binary sequence of period* $N = 4(2^m - 1)$, $m = 2k$, $k > 1$ *given by*

$$\mathbf{u} = \mathbf{a} + \mathbf{b} + \mathbf{c} \text{ where } u_i = a_i + b_i + c_i, \quad 0 \leq i \leq N - 1.$$

Then, $\mathbf{u}$ *has the optimal periodic autocorrelation magnitude, i.e.,* $C_\mathbf{u} \in \{N, 0, \pm 4\}$.

The sequence $\mathbf{u}$ has two different balancedness and autocorrelation distributions according to $\mathbf{z}$.

Theorem 1. *[30] Let* $\mathbf{u}$ *be a binary sequence generated by* $(z_0, z_1, z_2, z_3) = (1, 1, 0, 0)$ *or* $(1, 0, 0, 1)$ *in Application 2. Then, if the number of 1's in a period of* $\mathbf{u}$ *is denoted by* w, *then*

$$w = \frac{N}{2} - 1.$$

Thus, $\mathbf{u}$ *is almost balanced [20]. Also, the complete distribution of the periodic autocorrelation* $C_\mathbf{u}(\tau)$ *is given by*

$$C_\mathbf{u}(\tau) = \begin{cases} 4(2^m - 1), & 1 \text{ time} \\ 0, & 2^{2k} + 2^{k+1} - 3 \text{ times} \\ -4, & 2^{2k+1} - 2^k - 2 \text{ times} \\ +4, & 2^{2k} - 2^k \text{ times}. \end{cases}$$

Theorem 2. *[31] Let* $\mathbf{u}$ *be a binary sequence generated by* $(z_0, z_1, z_2, z_3) = (0, 0, 1, 1)$ *or* $(0, 1, 1, 0)$ *in Application 2. Then, if the number of 1's in a period of* $\mathbf{u}$ *is denoted by* w, *then*

$$w = \frac{N}{2} - \sqrt{N + 4} + 1.$$

Also, the complete distribution of the periodic autocorrelation $C_\mathbf{u}(\tau)$ *is given by*

$$C_\mathbf{u}(\tau) = \begin{cases} 4(2^m - 1), & 1 \text{ time} \\ 0, & 2^{2k} + 2^{k+1} - 3 \text{ times} \\ -4, & 2^k - 2 \text{ times} \\ +4, & 3 \cdot 2^k \cdot (2^k - 1) \text{ times}. \end{cases}$$

For more details on the autocorrelation and the linear feedback shift register (LFSR) configuration of the sequences, see [30] and [31].

4   The Perfect Sequence for Binary Sequences with Good Aperiodic Autocorrelation

Using the negacyclically perfect sequence '00', Parker has obtained binary sequences with merit factor 6.0 by applying a product method [22] [23]. In this section, we discuss the asymptotic merit factors of product sequences of the perfect sequence '0111' and other sequences with high merit factors. Also, we study the extensions of some of the product sequences by employing the method used in BCJ-KP sequences [8] [18]. Through intensive experiments, we obtain strong numerical evidences showing that the sequences have high asymptotic merit factors even if the proofs cannot yet be provided. Since the perfect sequence has length 4, the product sequences provide a large class of binary sequences of even lengths with high merit factors. From private communications, we noticed that all of these numerical results have been independently observed by Parker and Jedwab [24].

Before discussing the results, we define the *rotation* and the *appending* of a binary sequence.

Definition 2. *[8] Let* $\mathbf{a} = \{a_i\}$ *be a binary sequence of length* N. *Let* r *and* s *be real numbers with* $0 \le r \le 1$ *and* $0 \le s \le 1$. *Then, an* r-rotated *sequence* $\mathbf{a}_r$ *is defined by a* $\lfloor rN \rfloor$-*cyclic shift of* $\mathbf{a}$, *i.e.,*

$$\mathbf{a}_r = \{a_{i+\lfloor rN \rfloor} \mid 0 \le i \le N - 1\}.$$

Meanwhile, a t-appended *sequence* $\mathbf{a}^{(t)}$ *is defined by the appending of* $\lfloor tN \rfloor$ *elements to* $\mathbf{a}$, *i.e.,*

$$\mathbf{a}^{(t)} = \{a_i^{(t)} \mid 0 \le i \le N + \lfloor tN \rfloor - 1\}$$

$$\text{where } a_i^{(t)} = \begin{cases} a_i, & 0 \le i \le N - 1, \\ a_{i-N}, & N \le i \le N + \lfloor tN \rfloor - 1. \end{cases}$$

Therefore, an r-rotated *and* t-appended *sequence of* $\mathbf{a}$ *is defined by*

$$\mathbf{a}_r^{(t)} = \{a_{r,i}^{(t)} \mid 0 \le i \le N + \lfloor tN \rfloor - 1\}$$

$$\text{where } a_{r,i}^{(t)} = \begin{cases} a_{i+\lfloor rN \rfloor}, & 0 \le i \le N - 1, \\ a_{i+\lfloor rN \rfloor - N}, & N \le i \le N + \lfloor tN \rfloor - 1. \end{cases}$$

4.1   The Legendre and the Perfect Sequences

The Legendre sequence $\mathbf{a}$ of a prime period p is defined by

$$a_i = \left(\frac{i}{p}\right) = \begin{cases} 0, & \text{if } i = 0 \\ 0, & \text{if } i \text{ is quadratic residue modulo } p \\ 1, & \text{if } i \text{ is quadratic non-residue modulo } p \end{cases}$$

where $\left(\frac{i}{p}\right)$ is called the *Legendre symbol*. If $p \equiv 3 \pmod 4$, the Legendre sequence has the ideal two-level periodic autocorrelation corresponding to the

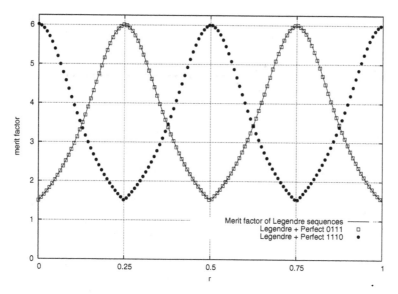

Fig. 1. Merit factors of r-rotated product sequences of length 12476

Paley-Hadamard matrix [21]. We assume $p \equiv 3 \pmod 4$ for all Legendre sequences employed in this paper. The asymptotic merit factor of the Legendre sequence has first been observed by Turyn [11], and theoretically proven by Høholdt and Jensen [14]. Let $\mathbf{a}_r$ be an r-rotated Legendre sequence of length p. As p goes to infinity, the asymptotic merit factor of the r-rotated Legendre sequence is given by [14]

$$\frac{1}{F(\mathbf{a}_r)} = \begin{cases} 8(r - \frac{1}{4})^2 + \frac{1}{6}, & \text{if } 0 \leq r \leq \frac{1}{2} \\ 8(r - \frac{3}{4})^2 + \frac{1}{6}, & \text{if } \frac{1}{2} \leq r \leq 1. \end{cases} \tag{2}$$

Hence, the maximum asymptotic merit factor of the Legendre sequence is 6 at $r = \frac{1}{4}$ and $\frac{3}{4}$.

Let $\mathbf{p} = \mathbf{a} + \mathbf{b}$ be a product sequence of length $N = 4p$, where $\mathbf{a}$ is the Legendre sequence of period p and $\mathbf{b}$ is the perfect binary sequence of period 4. Fig. 1 shows the merit factors of $\mathbf{p}_r$ with length $N = 12476$ at every rotation r, $0 \leq r \leq 1$. In Fig. 1, the solid line corresponds to the merit factors of Legendre sequences computed by (2). From the figure, we observed that $\mathbf{p}_r$ with $\mathbf{b} = (0, 1, 1, 1)$ has the same merit factors as Legendre sequences. Thus, we could assume that the maximum merit factors of $\mathbf{p}_r$ are also obtained at $r = \frac{1}{4}$ or $\frac{3}{4}$. If we apply the perfect sequence $\mathbf{b} = (1, 1, 1, 0)$ instead, then we obtain $\frac{N}{4}$-phase shifted merit factors of the corresponding $\mathbf{p}_r$, which can be easily understood.

Fig. 2 shows the merit factors of $\mathbf{p}_r$ of various lengths $N = 4p$. In Fig. 2, if $N < 20000$, then the merit factors are maximum over every rotations r. Otherwise, the merit factors are observed at $r = \frac{1}{4}$. From Fig. 2, numerical evidence suggests that the asymptotic merit factors of the product sequences of Legendre and perfect sequences are 6 at optimal rotations.

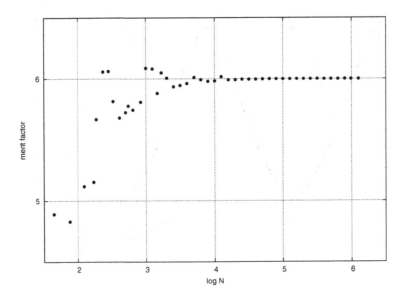

Fig. 2. Merit factors of product sequences (Legendre + Perfect)

Conjecture 1. *Let* $\mathbf{p} = \mathbf{a} + \mathbf{b}$ *be a product sequence of length* $N = 4p$, *where* $\mathbf{a}$ *is the Legendre sequence of period* p *and* $\mathbf{b} = (0, 1, 1, 1)$ *is the perfect binary sequence of period 4. Then, the* r*-rotated product sequence* $\mathbf{p}_r$ *has the asymptotic merit factor 6 at* $r = \frac{1}{4}$ *as* N *goes to infinity.*

If the Legendre sequence is replaced by the modified Jacobi sequence in Conjecture 1, then we have obtained similar numerical results of merit factors of 6 in [31], so we establish the following conjecture.

Conjecture 2. *Let* $\mathbf{p} = \mathbf{a} + \mathbf{b}$ *be a product sequence of length* $N = 4pq$, *where* $\mathbf{a}$ *is the modified Jacobi sequence of period* pq *for distinct primes* p *and* q, *and* $\mathbf{b} = (0, 1, 1, 1)$ *is the perfect binary sequence of period 4. Then, the* r*-rotated product sequence* $\mathbf{p}_r$ *has the asymptotic merit factor 6 at* $r = \frac{1}{4}$ *as* N *goes to infinity.*

4.2   The BCJ-KP Sequences and the Perfect Sequence

In [8], Borwein, Choi, and Jedwab presented new binary sequences with merit factors greater than 6.34. At the same time, Kristiansen and Parker also presented the equivalent binary sequences with different construction method [18]. The sequences are in fact r-rotated and t-appended Legendre sequences $\mathbf{a}_r^{(t)}$ given by Definition 2, where $\mathbf{a}$ is the Legendre sequence. Since their contributions are equivalently high to this research area, we call the sequences *BCJ-KP* sequences. In the BCJ-KP sequences, the highest merit factors are obtained at the optimal values of $r \approx 0.22$ and $t \approx 0.06$. Even if they couldn't provide

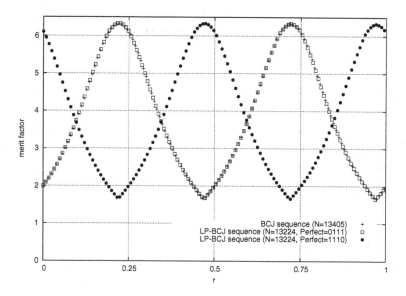

Fig. 3. Merit factors of r-rotated BCJ-KP and LP-BCJ-KP sequences ($t = 0.06$)

complete mathematical proofs, the results were strongly supported by extensive numerical observations.

Let $\mathbf{p} = \mathbf{a} + \mathbf{b}$ be a product sequence of length $4p$, where $\mathbf{a}$ is the Legendre sequence of a prime period p and $\mathbf{b}$ is the perfect binary sequence of period 4. In this paper, we construct an *LP-BCJ-KP sequence* $\mathbf{p}_r^{(t)}$ of length $N = 4p + \lfloor 4pt \rfloor$ by r-rotating and t-appending of $\mathbf{p}$, where 'LP' means *Legendre-Perfect*. Fig. 3 shows the merit factors of BCJ-KP and LP-BCJ-KP sequences – $\mathbf{a}_r^{(t)}$ and $\mathbf{p}_r^{(t)}$ – with given $t = 0.06$. Lengths of the sequences are 13405 and 13224, respectively. In Fig. 3, we see the merit factors of LP-BCJ-KP sequences $\mathbf{p}_r^{(t)}$ with $\mathbf{b} = (0, 1, 1, 1)$ are almost identical to those of BCJ-KP sequences $\mathbf{a}_r^{(t)}$ at all r's. Similar to Fig. 1, the perfect sequence $\mathbf{b} = (1, 1, 1, 0)$ generates $\mathbf{p}_r^{(t)}$ of length N whose merit factor is a $\frac{N}{4}$-cyclic shift of that of $\mathbf{p}_r^{(t)}$ with $\mathbf{b} = (0, 1, 1, 1)$. From Fig. 3, we assume that the LP-BCJ-KP sequences have the maximum merit factor of 6.34 at $r = 0.22$ with given $t = 0.06$.

Fig. 4 shows the merit factors of LP-BCJ-KP sequences $\mathbf{p}_r^{(t)}$ of various lengths N with $r = 0.22$ and $t = 0.06$. From the figure, we observed that the LP-BCJ-KP sequences have the asymptotic merit factors of 6.34.

Conjecture 3. *Let $\mathbf{p} = \mathbf{a} + \mathbf{b}$ be a product sequence of length $4p$, where $\mathbf{a}$ is the Legendre sequence of a prime period p and $\mathbf{b} = (0, 1, 1, 1)$ is the perfect binary sequence of period 4. Then, the LP-BCJ-KP sequence $\mathbf{p}_r^{(t)}$ of length $N = 4p + \lfloor 4pt \rfloor$ has the asymptotic merit factor 6.34 at $r \approx 0.22$ and $t \approx 0.06$ as N goes to infinity.*

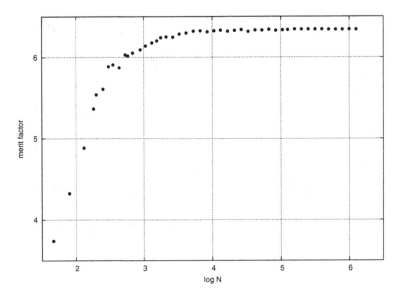

Fig. 4. Merit factors of LP-BCJ-KP sequences ($r = 0.22$ and $t = 0.06$)

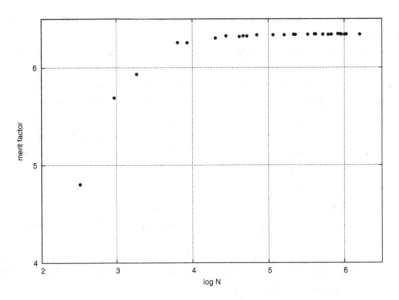

Fig. 5. Merit factors of MJP-BCJ-KP sequences ($r = 0.22$ and $t = 0.06$)

In [8], Borwein, Choi, and Jedwab pointed out that their approach – rotation and appending – could be also applied to (modified) Jacobi sequences. Similarly, modified Jacobi sequences could replace Legendre sequences in $\mathbf{p}_r^{(t)}$, which is called *MJP-BCJ-KP* sequence. Fig. 5 shows the merit factors of the MJP-BCJ-KP sequences of length $N = p(p + 4)$ with $r = 0.22$ and $t = 0.06$. From the numerical results, we can also conjecture that the asymptotic merit factors of the MJP-BCJ-KP sequences are 6.34.

Conjecture 4. *Let* $\mathbf{p} = \mathbf{a} + \mathbf{b}$ *be a product sequence of length* $4p(p + 4)$, *where* $\mathbf{a}$ *is the modified Jacobi sequence of period* $p(p + 4)$ *and* $\mathbf{b} = (0, 1, 1, 1)$ *is the perfect binary sequence of period 4. Then, the MJP-BCJ-KP sequence* $\mathbf{p}_r^{(t)}$ *of length* $N = 4p(p + 4) + \lfloor 4pt(p + 4) \rfloor$ *has the asymptotic merit factor 6.34 at* $r \approx 0.22$ *and* $t \approx 0.06$ *as* N *goes to infinity.*

5   Conclusion

We have investigated the effectiveness of the perfect binary sequence in applications for low periodic and aperiodic autocorrelations.

First, the perfect sequence could be employed for generating binary sequences with optimal periodic autocorrelation by means of a product method and its slight modification. The application is supported by theoretical proofs.

Second, the perfect sequence could be also applied for extending periods of binary sequences by a product method, where high merit factors of original sequences are preserved in the product sequences. If we further apply the rotation and appending to some of the product sequences, the resulting sequences have merit factors 6.34 like BCJ-KP sequences [8] [18].

Acknowledgement

The authors would like to thank Dr. Matthew G. Parker and Professor Jonathan Jedwab for their valuable comments on Section 4.

References

1. Physical Layer Standard for cdma2000 Spread Spectrum Systems, Ver. 1.0, Revision D (February 2004)
2. 3GPP TS 25.213 v.3.9.0, Technical Specification Group Radio Access Network; Spreading and modulation (FDD) (Release 1999) (December 2003)
3. IEEE Standard 802.11, Information Technology - Telecommunications and Information Exchange between Systems - Local and Metropolitan Area Networks - Specific Requirements Part 11: Wireless LAN Medium Access Control (MAC) and Physical Layer (PHY) Specifications (2003)
4. Arasu, K.T., Ding, C., Helleseth, T., Kumar, P.V., Martinsen, H.: Almost difference sets and their sequences with optimal autocorrelation. IEEE Trans. Inform. Theory 47(7), 2934–2943 (2001)
5. Barker, R.H.: Group synchronization of binary digital systems. In: Proceedings of the Second London Symposium on Information Theory, pp. 273–287 (1953)
6. Baumert, L.D.: Cyclic Difference Sets. Springer, Berlin (1971)

7. Beenker, G.F.M., Claasen, T.A.C.M., Hermens, P.W.C.: Binary sequences with a maximally flat amplitude spectrum. Philips J. Res 40, 289–304 (1985)

8. Borwein, P., Choi, K.-K.S., Jedwab, J.: Binary sequences with merit factor greater than 6.34. IEEE Inform. Theory 50, 3234–3249 (2004)

9. Dillon, J.F., Dobbertin, H.: New cyclic difference sets with Singer parameters. Finite Fields and Their Applications 10, 342–389 (2004)

10. Golay, M.J.E.: Sieves for low autocorrelation binary sequences. IEEE Trans. Inform. Theory IT-23, 43–51 (1977)

11. Golay, M.J.E.: The merit factor of Legendre sequences. IEEE Trans. Inform. Theory IT-29, 934–936 (1983)

12. Golomb, S.W., Gong, G.: Signal Design for Good Correlation - for Wireless Communication, Cryptography and Radar. Cambridge University Press, Cambridge (2005)

13. Hall, M.: A survey of difference sets. Proc. Amer. Math. Soc. 7, 975–986 (1956)

14. Høholdt, T., Jensen, H.E.: Determination of the merit factor of Legendre sequences. Trans. Inform. Theory 34(1), 161–164 (1988)

15. Jedwab, J.: A survey of the merit factor problem for binary sequences. In: Helleseth, T., Sarwate, D., Song, H.-Y., Yang, K. (eds.) SETA 2004. LNCS, vol. 3486, pp. 30–55. Springer, Heidelberg (2005)

16. Jungnickel, D., Pott, A.: Difference sets: An introduction. In: Pott, A., Kumar, P.V., Helleseth, T., Jungnickel, D. (eds.) Difference Sets, Sequences and their Correlation Properties. NATO Science Series C, vol. 542, pp. 259–296 (1999)

17. Kristiansen, R.: On the Aperiodic Autocorrelation of Binary Sequences, Master's Thesis, Department of Informatics, University of Bergen (2003)

18. Kristiansen, R., Parker, M.G.: Binary sequences with merit factor > 6.3. IEEE Trans. Inform. Theory. 50, 3385–3389 (2004)

19. Lüke, H.D.: Sequences and arrays with perfect periodic correlation. IEEE Trans. Aerosp. Electron. Syst. 24(3), 287–294 (1988)

20. No, J.-S., Chung, H., Song, H.-Y., Yang, K., Lee, J.-D., Helleseth, T.: New construction for binary sequences of period $p^m - 1$ with optimal autocorrelation using $(z + 1)^d + az^d + b$. IEEE Trans. Inform. Theory 47(4), 1638–1644 (2001)

21. Paley, R.E.A.C.: On orthogonal matrices. J. Math. Phys. 12, 311–320 (1933)

22. Parker, M.G.: Even length binary sequence families with low negaperiodic autocorrelation. In: Bozta, S., Sphparlinski, I. (eds.) Applied Algebra, Algebraic Algorithms and Error-Correcting Codes. LNCS, vol. 2227, pp. 200–209. Springer, Heidelberg (2001)

23. Parker, M.G.: Univariate and multivariate merit factors. In: Helleseth, T., Sarwate, D., Song, H.-Y., Yang, K. (eds.) SETA 2004. LNCS, vol. 3486, pp. 72–100. Springer, Heidelberg (2005)

24. Parker, M.G., Jedwab, J.: Private Communications (2007)

25. Paterson, K.G.: Applications of exponential sums in communication theory. In: Walker, M. (ed.) Cryptography and Coding. LNCS, vol. 1746, pp. 1–24. Springer, Heidelberg (1999)

26. Ireland, K., Rosen, M.: A Classical Introduction to Modern Number theory, 2nd edn. Springer, New York (1991)

27. Schmidt, B.: Cyclotomic integers and finite geometry. J. Amer. Math. Soc. 12, 929–952 (1999)

28. Scholtz, R.A., Pozar, D.M., Namgoong, W.: Ultra-Wideband Radio. EURASIP Journal on Applied Signal Processing 3, 252–272 (2005)
29. Turyn, R.J., Storer, J.: On binary sequences. Proc. Amer. Math. Soc. 12, 394–399 (1961)
30. Yu., N.Y., Gong, G.: New binary sequences with optimal autocorrelation magnitude. IEEE Trans. Inform. Theory (submitted to) (August 2006)
    http://www.cacr.math.uwaterloo.ca/techreports/2006/cacr2006-29.pdf
31. Yu, N.Y., Gong, G.: Applications of the perfect binary sequence of period 4 for low periodic and aperiodic autocorrelations. In: Golomb, S.W., Gong, G., Helleseth, T., Song, H.-Y. (eds.) SEQUENCES 2007. LNCS, vol. 4893, pp. 22–43. Springer, Heidelberg (2007)

On the Dual of Monomial Quadratic p-ary Bent Functions*

Tor Helleseth and Alexander Kholosha

The Selmer Center
Department of Informatics, University of Bergen
P.O. Box 7800, N-5020 Bergen, Norway
{Tor.Helleseth,Alexander.Kholosha}@uib.no

Abstract. Considered are quadratic p-ary bent functions having the form $f(x) = \mathrm{Tr}_n(ax^{p^j+1})$. Described is the general Gold-like class of bent functions that covers all the previously known monomial quadratic cases. Obtained is the exact value of the Walsh transform coefficients for a bent function in this class. In particular, presented is an explicit expressions for a dual of a monomial quadratic bent function which is a bent functions on its own. This gives new examples of generalized bent functions not previously reported in the literature. The paper is the follow-up to Helleseth-Kholosha 2006.

Keywords: p-ary bent function, regularity, dual function, exponential sum.

1   Introduction

Boolean bent functions were first introduced by Rothaus in 1976 as an interesting combinatorial object with the important property of having the maximum Hamming distance to the set of all affine functions. Later the research in this area was stimulated by the significant relation to the following topics in computer science: coding theory, sequences and cryptography (design of stream ciphers and S-boxes for block ciphers). Kumar, Scholtz and Welch in [1] generalized the notion of Boolean bent functions to the case of functions over an arbitrary finite field. Complete classification of bent functions looks hopeless even in the binary case. In the case of generalized bent functions things are naturally much more complicated. However, many explicit methods are proved for constructing bent functions either from scratch or based on other, simpler bent functions.

Given a function $f(x)$ mapping $\mathrm{GF}(p^n)$ to $\mathrm{GF}(p)$, the direct and inverse *Walsh transform* operations on f are defined at a point by the following respective identities:

$$S_f(b) = \sum_{x \in \mathrm{GF}(p^n)} \omega^{f(x) - \mathrm{Tr}_n(bx)} \quad \text{and} \quad \omega^{f(x)} = \frac{1}{p^n} \sum_{b \in \mathrm{GF}(p^n)} S_f(b) \omega^{\mathrm{Tr}_n(bx)}$$

---

* This work was supported by the Norwegian Research Council.

S.W. Golomb et al. (Eds.): SSC 2007, LNCS 4893, pp. 50–61, 2007.

where $\mathrm{Tr}_n() : \mathrm{GF}(p^n) \rightarrow \mathrm{GF}(p)$ denotes the absolute trace function, $\omega = e^{\frac{2\pi i}{p}}$ is the complex primitive p^{th} root of unity and elements of $\mathrm{GF}(p)$ are considered as integers modulo p. In the sequel, $S_a(b)$ is also used to denote the Walsh transform coefficient of a function that depends on parameter a when it is clear from the context which function we mean.

According to [1], $f(x)$ is called a *p-ary bent function* (or *generalized bent function*) if all its Walsh coefficients satisfy $|S_f(b)|^2 = p^n$. A bent function $f(x)$ is called *regular* (see [1, Definition 3] and [2, p. 576]) if for every $b \in \mathrm{GF}(p^n)$ the normalized Walsh coefficient $p^{-n/2}S_f(b)$ is equal to a complex p^{th} root of unity, i.e., $p^{-n/2}S_f(b) = \omega^{f^*(b)}$ for some function f^* mapping $\mathrm{GF}(p^n)$ into $\mathrm{GF}(p)$. A bent function $f(x)$ is called *weakly regular* if there exists a complex u having unit magnitude such that $up^{-n/2}S_f(b) = \omega^{f^*(b)}$ for all $b \in \mathrm{GF}(p^n)$. We call $u^{-1}p^{n/2}$ the *magnitude* of $S_f(b)$. Throughout this paper, $p^{n/2}$ with odd n stands for the *positive* square root of p^n. A function $F(x)$ mapping $\mathrm{GF}(p^n)$ to itself will also be called generalized bent if $\mathrm{Tr}_n(F(x))$ is bent according to the above definition. In the present paper, we take an odd prime p and examine prospective *monomial* quadratic p-ary bent functions having the form $f(x) = \mathrm{Tr}_n\left(ax^{p^j+1}\right)$ with $a, x \in \mathrm{GF}(p^n)$, $a \neq 0$.

Weakly regular bent functions always appear in pairs. Indeed, if $f(x)$ is a (weakly) regular bent function and $S_f(b) = u^{-1}p^{n/2}\omega^{f^*(b)}$ for $b \in \mathrm{GF}(p^n)$, then the function $f^*(b)$ is called the *dual* of f. The inverse Walsh transform of such $f(x)$ gives

$$up^{n/2}\omega^{f(x)} = \sum_{b \in \mathrm{GF}(p^n)} \omega^{f^*(b)+\mathrm{Tr}_n(bx)} = S_{f^*}(-x) \ .$$

Thus, the dual of a (weakly) regular bent function is again a (weakly) regular bent function and $f^{**}(x) = f(-x)$, $f^{***}(x) = f^*(-x)$, $f^{****}(x) = f(x)$.

Following the definition of a bent function, the standard method for proving that a function is bent would be to evaluate the absolute square of its Walsh coefficients. However, this technique does not help in telling if the function is (weakly) regular and in finding its dual. In the current paper, we evaluate explicitly the Walsh transform coefficients of monomial quadratic bent functions over finite fields of odd characteristic. This is the main contribution here if compared to a recent paper [3] by the same authors.

There are only a few proven cases of monomial bent functions and most of them are quadratic. For a comprehensive reference on this topic we refer the reader to [3]. In this paper, we prove the general Gold-like form of a monomial quadratic bent function that covers the known Sidelnikov, Kumar-Moreno [4] and Kasami [5] cases. The Walsh transform coefficients for Sidelnikov and Kumar-Moreno bent functions were explicitly calculated by Coulter in [6, Theorem 1 (i), Lemma 3.2]. Here, we solve this problem completely for any monomial quadratic bent function and show that any such a function is (weakly) regular. In particular, we find dual functions that give new examples of generalized bent functions not previously reported in the literature.

2 Preliminaries

We start with the following lemmas. Let $v(b)$ denote the (additive) 2-adic valuation of integer b (i.e., the maximal power of 2 dividing b). Let ξ be a primitive element of $\mathrm{GF}(p^n)$ and for $a \in \mathrm{GF}(p^n)^*$ define $\mathrm{ind}(a)$ as the unique integer t with $a = \xi^t$ and $0 \leq t < p^n - 1$. According to [7, Example 5.10], the real-valued function η on $\mathrm{GF}(p^n)^*$ with $\eta(c) = 1$ if c is the square of an element of $\mathrm{GF}(p^n)^*$ and $\eta(c) = -1$ otherwise, is called the *quadratic character* of $\mathrm{GF}(p^n)$. The following lemma can also be found in [8, Lemma 2.6] and [9, Lemma 7.2].

Lemma 1. *For an odd prime p*

$$\gcd\left(p^j + 1, p^n - 1\right) = \begin{cases} p^{\gcd(j,n)} + 1, & \text{if } v(j) < v(n) \\ 2, & \text{otherwise} . \end{cases}$$

Proof. Denote $d = \gcd(p^j+1, p^n-1)$. It is easy to see that if $v(p^{2j}-1) \leq v(p^n-1)$ then

$$p^{\gcd(2j,n)} - 1 = \gcd(p^{2j} - 1, p^n - 1) = d\gcd(p^j - 1, p^n - 1) = d(p^{\gcd(j,n)} - 1)$$

since $\gcd(p^j - 1, p^j + 1) = 2$. Alternatively, if $v(p^n - 1) < v(p^{2j} - 1)$ then

$$2(p^{\gcd(2j,n)} - 1) = d(p^{\gcd(j,n)} - 1) .$$

Note that if n is odd then $d = 2$. Further consider only even n.

It is well known that $v(p^b - 1) = v(p + 1) + v(b)$ if $p \equiv 3 \pmod 4$ and b is even; in the remaining cases $v(p^b - 1) = v(p - 1) + v(b)$. Thus, for even n the condition $v(p^n - 1) < v(p^{2j} - 1)$ is equivalent to $v(n) \leq v(j)$. In this case, $d = 2$ as well. Otherwise we have

$$d = \frac{p^{\gcd(2j,n)} - 1}{p^{\gcd(j,n)} - 1} = \frac{p^{2\gcd(j,n)} - 1}{p^{\gcd(j,n)} - 1} = p^{\gcd(j,n)} + 1$$

when $v(j) < v(n)$.   □

Lemma 2. *Let p be an odd prime, $j \in \{1,\ldots,n\}$ and $a \in \mathrm{GF}(p^n)$ be nonzero.*

(i) If $v(n) \leq v(j)$ then

$$\sum_{x \in \mathrm{GF}(p^n)} \omega^{\mathrm{Tr}_n\left(ax^{p^j+1}\right)} = \begin{cases} \eta(a)(-1)^{n-1}p^{n/2}, & \text{if } p \equiv 1 \pmod 4 \\ \eta(a)(-1)^{n-1}i^n p^{n/2}, & \text{if } p \equiv 3 \pmod 4 . \end{cases}$$

(ii) If $v(n) > v(j) + 1$ then

$$\sum_{x \in \mathrm{GF}(p^n)} \omega^{\mathrm{Tr}_n\left(ax^{p^j+1}\right)} = \begin{cases} -p^{n/2+\gcd(j,n)}, & \text{if } a \in C_0 \\ p^{n/2}, & \text{otherwise} . \end{cases}$$

(iii) If $v(n) = v(j) + 1$ then

$$\sum_{x \in \mathrm{GF}(p^n)} \omega^{\mathrm{Tr}_n\left(ax^{p^j+1}\right)} = \begin{cases} p^{n/2+\gcd(j,n)}, & \text{if } a \in C_{d/2} \\ -p^{n/2}, & \text{otherwise} \end{cases}$$

where $d = p^{\gcd(j,n)} + 1$ and $C_t = \{a \in \mathrm{GF}(p^n)^* \mid \mathrm{ind}(a) \equiv t \pmod d\}$.

Proof. First, note that if ξ is a primitive element of $\mathrm{GF}(p^n)$ then

$$\sum_{x \in \mathrm{GF}(p^n)} \omega^{\mathrm{Tr}_n\left(ax^{p^j+1}\right)} = 1 + \sum_{t=0}^{p^n-2} \omega^{\mathrm{Tr}_n\left(a\xi^{t(p^j+1)}\right)}$$

$$= 1 + \sum_{t=0}^{p^n-2} \omega^{\mathrm{Tr}_n\left(a\xi^{td}\right)} = \sum_{x \in \mathrm{GF}(p^n)} \omega^{\mathrm{Tr}_n\left(ax^d\right)}$$

where $d = \gcd(p^j + 1, p^n - 1)$ and since

$$\{t(p^j+1) \pmod{p^n-1} \mid t = 0, \ldots, p^n-2\} = \{td \pmod{p^n-1} \mid t = 0, \ldots, p^n-2\}.$$

Therefore, the exponents $p^j + 1$ can be replaced with d without affecting the value of the questioned sum. The value of d is given by Lemma 1. For those values of j giving $d = 2$ we use [7, Theorems 5.15, 5.33] and the remaining cases when $v(j) < v(n)$ are settled by [10, Lemma 3.5] (see [11,8] for the proofs). $\square$

Lemma 3. *For any nonzero $a \in \mathrm{GF}(p^n)$ and $j \in \{1, \ldots, n\}$ assume $e = n/\gcd(j,n)$ is even. Let $d = p^{\gcd(j,n)} + 1$ and*

$$D(a) = (-1)^{\frac{e}{2}} \left(a^{\frac{p^{je}-1}{p^j+1}} + a^{-\frac{p^{je}-1}{p^j+1}} \right) - 2 . \tag{1}$$

Then $D(a) \in \mathrm{GF}(p^j)$ and $D(a) = 0$ if and only if one of the following holds

(i) $v(n) > v(j) + 1$ and d divides i_0,
(ii) $v(n) = v(j) + 1$ and $d/2$ divides i_0 with $2i_0/d$ being odd,

where $i_0 = \mathrm{ind}(a)$.

Proof. Obviously, $\frac{p^j(p^{je}-1)}{p^j+1} \equiv -\frac{p^{je}-1}{p^j+1} \pmod{p^n-1}$ and, thus, $D(a)^{p^j} = D(a)$ so $D(a) \in \mathrm{GF}(p^j)$. It is also clear that $D(a) = 0$ can be viewed as a quadratic equation having a unique solution $a^{\frac{p^{je}-1}{p^j+1}} = (-1)^{\frac{e}{2}}$. The latter holds if and only if

$$\frac{i_0(p^{je}-1)}{p^j+1} \equiv \frac{e(p^n-1)}{4} \pmod{p^n-1} .$$

Note that p^n-1 divides $p^{je}-1$ since n divides je and that $\gcd(p^j+1, p^n-1) = d$, by Lemma 1, since e is even that is equivalent to $v(n) > v(j)$. Therefore, all the terms in the latest equivalence can be divided by $\frac{p^n-1}{d}$ which leads to

$$\frac{i_0(p^{je}-1)d}{(p^n-1)(p^j+1)} \equiv \frac{ed}{4} \pmod d . \tag{2}$$

Further, if $\gcd\left(d, \frac{p^{je}-1}{p^j+1}\right) = q_1$ then $p^{\gcd(j,n)} \pmod{q_1} \equiv -1$ and

$$p^{j(e-i)} = \left(p^{\gcd(j,n)}\right)^{\frac{j}{\gcd(j,n)}(e-i)} \equiv (-1)^{e-i} \pmod{q_1}$$

for any $i = \{1, \ldots, e\}$. Thus, since e is even,

$$\frac{p^{je}-1}{p^j+1} = p^{j(e-1)} - p^{j(e-2)} + \cdots - 1 \equiv -e \equiv 0 \pmod{q_1} \ .$$

In a similar way, it can be shown that if $\gcd\left(d, \frac{p^{je}-1}{p^n-1}\right) = q_2$ then $\frac{j}{\gcd(j,n)} = \frac{je}{n} \equiv 0 \pmod{q_2}$. Now we conclude that

$$\gcd\left(d, \frac{\left(p^{je}-1\right)d}{(p^n-1)\left(p^j+1\right)}\right) \bigg| \gcd(q_1, q_2) \bigg| \gcd\left(\frac{n}{\gcd(j,n)}, \frac{j}{\gcd(j,n)}\right) = 1 \ .$$

Thus, d is coprime to $\frac{\left(p^{je}-1\right)d}{(p^n-1)(p^j+1)}$.

(i) First, assume $v(n) > v(j) + 1$ which gives $v(e) = v(n) - v(j) > 1$ meaning that 4 divides e. In this case, (2) holds if and only if d divides $\frac{i_0\left(p^{je}-1\right)d}{(p^n-1)(p^j+1)}$ that is equivalent to d divides i_0.

(ii) Now assume $v(n) = v(j) + 1$ which gives $v(e) = v(n) - v(j) = 1$. In this case, (2) holds if and only if $d/2$ divides $\frac{i_0\left(p^{je}-1\right)d}{(p^n-1)(p^j+1)}$ and $\frac{2i_0\left(p^{je}-1\right)}{(p^n-1)(p^j+1)} - \frac{e}{2}$ is even. The first of these two conditions is equivalent to $d/2$ divides i_0 and, thus, the second condition means $\frac{d(2i_0/d)\left(p^{je}-1\right)}{(p^n-1)(p^j+1)}$ is odd, since $e/2$ is odd. The latter holds if and only if

$$v(d) + v(2i_0/d) + v(p^{je}-1) = v(p^n-1) + v(p^j+1)$$

which becomes $v(d) + v(2i_0/d) = v(p^j+1)$, using the identities for $v(p^b-1)$ from the proof of Lemma 1. Now, using the same identities and $v(p^{2b}-1) = v(p^b-1)+v(p^b+1)$, it can be easily obtained that $v(p^b+1) = v(p+1)-v(p-1)+1$ if $p \equiv 3 \pmod 4$ and b is odd; in the remaining cases $v(p^b+1) = 1$. Therefore, $v(d) = v(p^{\gcd(j,n)}+1) = v(p^j+1)$ and $v(2i_0/d) = 0$ which is exactly our claim that $2i_0/d$ is odd. $\qquad\square$

3   Quadratic Monomial Bent Functions

In this section, we consider quadratic monomial functions with the exponent of the Gold type. Our approach allows to derive explicit requirements on the value of the coefficient a. Moreover, these results provide the generalization for the known monomial cases of p-ary bent functions due to Sidelnikov, Kumar-Moreno and Kasami. We also study closely the property of these bent functions to be (weakly) regular. To that end, we prove that any monomial quadratic bent function is (weakly) regular and give the exact value of the Walsh transform coefficients for a bent function in this class. In particular, we obtain an expression for the dual of such functions.

Take a quadratic function $F(x) = \sum_{i=0}^{n-1} a_i x^{p^i+1}$ and define

$$F^*(x) = \sum_{i=0}^{n-1} \left(a_i^{p^I} x^{p^{I+i}} + a_i^{p^{I-i}} x^{p^{I-i}} \right) ,$$

where $I = \max\{i \mid a_i \neq 0;\ i = 0, 1, \ldots, n-1\}$. Note that $F^*(x)$ is a linearized polynomial. Then, according to [9, Corollary 3.2],

$$S_f(b) = \begin{cases} \omega^{-f(x_0)} S_f(0), & \text{if } F^*(x) = -b^{p^I} \text{ has a solution } x_0 \in \mathrm{GF}(p^n) , \\ 0, & \text{otherwise} , \end{cases} \quad (3)$$

where $f(x) = \mathrm{Tr}_n(F(x))$ (similar result for monomial quadratic functions was earlier proved in [6, Theorem 3.1]). Therefore, if $f(x)$ is bent then the equation $F^*(x) = -b^{p^I}$ has a solution for any $b \in \mathrm{GF}(p^n)$ (this solution is unique) or, equivalently, $F^*(x) = 0$ only for $x = 0$ (since if the linear operator on $\mathrm{GF}(p^n)$ defined by $F^*(x)$ has the kernel of dimension zero then the image contains all the elements of $\mathrm{GF}(p^n)$). It was also shown in [9, Lemmas 2.1, 3.1] that $|S_f(0)| = p^{(2n-r)/2}$, where r is the rank of the quadratic form associated with f and that the equation $F^*(x) = 0$ has p^{n-r} solutions. Thus, for the reverse implication, if $F^*(x) = 0$ only for $x = 0$ then $r = n$ and $f(x)$ is bent.

Theorem 1. *Let $a \in \mathrm{GF}(p^n)$ be nonzero and a prime p be odd. Then for any $j \in \{1, \ldots, n\}$, the quadratic p-ary function $f(x)$ mapping $\mathrm{GF}(p^n)$ to $\mathrm{GF}(p)$ and given by*

$$f(x) = \mathrm{Tr}_n\left(ax^{p^j+1}\right) \quad (4)$$

is bent if and only if

$$p^{\gcd(2j,n)} - 1 \nmid \frac{p^n - 1}{2} - \mathrm{ind}(a)(p^j - 1) . \quad (5)$$

Moreover, if (5) holds then $f(x)$ is a (weakly) regular bent function and for $b \in \mathrm{GF}(p^n)$ the corresponding Walsh transform coefficient of $f(x)$ is equal to

$$S_a(b) = S_a(0)\omega^{-\mathrm{Tr}_n\left(ax_0^{p^j+1}\right)} , \quad (6)$$

where x_0 is a unique solution of the equation $a^{p^j} x^{p^{2j}} + ax = -b^{p^j}$. Further, if $e = n/\gcd(j,n)$ is odd then (5) is satisfied by any nonzero a and

$$x_0 = -\frac{1}{2} \sum_{t=0}^{e-1} (-1)^t a^{-\frac{p^j(2t+1)+1}{p^j+1}} b^{p^j(2t+1)} . \quad (7)$$

If e is even and (5) holds then $D(a)$ defined in (1) is nonzero and

$$D(a)x_0 = \sum_{t=0}^{e/2-1} (-1)^t \left((-1)^{\frac{e}{2}+1} a^{-\frac{p^j(2t+1)+p^je}{p^j+1}} + a^{-\frac{p^j(2t+1)+1}{p^j+1}} \right) b^{p^j(2t+1)} .$$

Finally, the magnitude of $S_a(b)$ that is equal to $S_a(0)$ can be determined using Lemma 2 given the concrete values of j and n.

Proof. By the arguments following (3), $f(x)$ of type (4) is bent if and only if $F^*(x) = a^{p^j} x^{p^{2j}} + ax = 0$ only for $x = 0$. If x is nonzero then $F^*(x) = 0$ is equivalent to $x^{p^{2j}-1} = -a^{1-p^j}$. Let $i_0 = \text{ind}(a) \in \{0, \ldots, p^n - 2\}$. Then $-a^{1-p^j} = \xi^{\frac{p^n-1}{2}-i_0(p^j-1)}$. Consider the following equation of the unknown $t \in \{0, \ldots, p^n - 2\}$

$$\xi^{\frac{p^n-1}{2}-i_0(p^j-1)} = \xi^{(p^{2j}-1)t}$$

which holds if and only if $\frac{p^n-1}{2} - i_0(p^j-1) \equiv (p^{2j}-1)t \pmod{(p^n-1)}$. The latter congruence has a solution in t if and only if

$$\gcd\left(p^{2j} - 1, p^n - 1\right) = p^{\gcd(2j,n)} - 1 \left| \frac{p^n-1}{2} - i_0(p^j-1)\right. .$$

Thus, condition (5) holds if and only if $a^{p^j} x^{p^{2j}} \neq -ax$ for any nonzero $x \in \text{GF}(p^n)$ which is equivalent to $f(x)$ being bent.

Identity (6) for $S_a(b)$ follows immediately from (3) and in its turn, by (6), any bent function having the form of (4) is (weakly) regular. Recall that for those $a \in \text{GF}(p^n)^*$ satisfying (5) we have $a^{p^j} x^{p^{2j}} + ax \neq 0$ unless $x = 0$. It means that this linear operator on $\text{GF}(p^n)$ has the kernel of dimension zero and thus, the image contains all the elements of $\text{GF}(p^n)$. Therefore, $a^{p^j} x^{p^{2j}} + ax = -b^{p^j}$ has a unique solution x_0 for any $b \in \text{GF}(p^n)$.

It can be checked directly that our expressions for x_0 are correct since if e is odd then

$$a^{p^j}(2x_0)^{p^{2j}} = -\sum_{t=0}^{e-1}(-1)^t a^{p^j - \frac{p^{2j}\left(p^{j(2t+1)}+1\right)}{p^j+1}} b^{p^{j(2t+3)}}$$

$$= -\sum_{t=0}^{e-1}(-1)^t a^{1-\frac{p^{j(2t+3)}+1}{p^j+1}} b^{p^{j(2t+3)}} = a\sum_{t=1}^{e}(-1)^t a^{-\frac{p^{j(2t+1)}+1}{p^j+1}} b^{p^{j(2t+1)}}$$

$$= -2ax_0 - b^{p^j} - a^{1-\frac{p^{j(2e+1)}+1}{p^j+1}} b^{p^j} = -2ax_0 - 2b^{p^j} ,$$

since

$$\frac{p^{j(2e+1)}+1}{p^j+1} = \sum_{t=0}^{2e}(-1)^t p^{jt} \equiv \sum_{t=0}^{e-1}(-1)^t p^{jt} + \sum_{t=0}^{e-1}(-1)^{t+1} p^{jt} + 1 \pmod{p^n - 1} .$$

We refer the reader also to [6, Lemma 3.2] where a direct method to solve the equation $= a^{p^j} x^{p^{2j}} + ax = -b^{p^j}$ with e odd is presented.

If e is even then

$$a^{p^j}(D(a)x_0)^{p^{2j}}$$

$$= \sum_{t=0}^{e/2-1}(-1)^t\left((-1)^{\frac{e}{2}+1}a^{p^j-\frac{p^{j(2t+3)}\left(p^{j(e-2t-1)}+1\right)}{p^j+1}} + a^{p^j-\frac{p^{2j}\left(p^{j(2t+1)}+1\right)}{p^j+1}}\right) b^{p^{j(2t+3)}}$$

$$= \sum_{t=0}^{e/2-2} (-1)^t \left((-1)^{\frac{e}{2}+1} a^{1 - \frac{p^{j(2t+3)}\left(p^{j(e-2t-3)}+1\right)}{p^j+1}} + a^{1 - \frac{p^{j(2t+3)}+1}{p^j+1}} \right) b^{p^{j(2t+3)}}$$

$$+ \left(1 + (-1)^{\frac{e}{2}-1} a^{1 - \frac{p^{j(e+1)}+1}{p^j+1}} \right) b^{p^{j(e+1)}}$$

$$= -a \sum_{t=1}^{e/2-1} (-1)^t \left((-1)^{\frac{e}{2}+1} a^{-\frac{p^{j(2t+1)}\left(p^{j(e-2t-1)}+1\right)}{p^j+1}} + a^{-\frac{p^{j(2t+1)}+1}{p^j+1}} \right) b^{p^{j(2t+1)}}$$

$$+ \left(1 - (-1)^{\frac{e}{2}} a^{\frac{p^{je}-1}{p^j+1}} \right) b^{p^j}$$

$$= -aD(a)x_0 + \left(2 - (-1)^{\frac{e}{2}} a^{\frac{p^{je}-1}{p^j+1}} - (-1)^{\frac{e}{2}} a^{-\frac{p^{je}-1}{p^j+1}} \right) b^{p^j}$$

$$= -aD(a)x_0 - D(a)b^{p^j} \ .$$

Now we have to prove that when e is even (i.e., $v(n) > v(j)$) then $D(a) \neq 0$ if (5) holds. In this case, $\gcd(2j, n) = 2 \gcd(j, n)$ and denote $d = p^{\gcd(j,n)} + 1$. First, assume $v(n) > v(j) + 1$. Since for any positive integers g and l holds

$$\frac{p^{gl} - 1}{p^g - 1} = p^{g(l-1)} + p^{g(l-2)} + \ldots + 1 \equiv l \pmod 2 \ , \tag{8}$$

we have $\frac{p^n - 1}{p^{\gcd(2j,n)} - 1} \equiv \frac{n}{\gcd(2j,n)} \equiv 0 \pmod 2$ and thus, $p^{\gcd(2j,n)} - 1$ divides $(p^n - 1)/2$. Therefore, (5) holds if and only if $p^{\gcd(2j,n)} - 1$ does not divide $i_0(p^j - 1)$ or, equivalently, d does not divide $i_0(p^j - 1)/(p^{\gcd(j,n)} - 1)$. Note that $v(j) = v(\gcd(j, n))$ and, by Lemma 1, $\gcd(d, p^j - 1) = 2$. Again, by (8), $\frac{p^j - 1}{p^{\gcd(j,n)} - 1} \equiv \frac{j}{\gcd(j,n)} \equiv 1 \pmod 2$ and thus, d is coprime to $(p^j - 1)/(p^{\gcd(j,n)} - 1)$ and (5) holds if and only if d does not divide i_0.

Finally, assume $v(n) = v(j) + 1$. By (8), both $(p^n - 1)/(p^{\gcd(2j,n)} - 1)$ and $(p^j - 1)/(p^{\gcd(j,n)} - 1)$ are odd. Similarly to the previous case, it can be proved that the condition

$$\frac{p^{\gcd(2j,n)} - 1}{2} \ \bigg| \ \frac{p^n - 1}{2} - i_0(p^j - 1)$$

holds if and only if $d/2$ divides i_0. Therefore, (5) does not hold if and only if $d/2$ divides i_0 and

$$\frac{p^n - 1}{p^{\gcd(2j,n)} - 1} - \frac{2i_0}{d} \frac{p^j - 1}{p^{\gcd(j,n)} - 1}$$

is even or, equivalently, $2i_0/d$ is odd. We conclude that if e is even and (5) holds then, by Lemma 3, $D(a) \neq 0$. Thus, $a^{p^j} x_0^{p^{2j}} + ax_0 = -b^{p^j}$ since $D(a) \in GF(p^j)^*$.

To finalize the proof, note that the value of the Walsh transform of $f(x)$ in point zero is equal to $S_a(0) = \sum_{x \in GF(p^n)} \omega^{\text{Tr}_n\left(ax^{p^j+1}\right)}$ and can be found using Lemma 2. Since f is a (weakly) regular bent function, the magnitude of $S_a(b)$ does not depend on b and is equal to the magnitude of $S_a(0)$. $\qquad \square$

Corollary 1. *Given the conditions of Theorem 1, $f(x) = \mathrm{Tr}_n\left(x^{p^j+1}\right)$ is a bent function if and only if $\frac{n}{\gcd(2j,n)}$ is odd. Moreover, if the latter condition holds then $f(x)$ is (weakly) regular.*

Proof. Consider function (4) with $a = 1$. This is a bent function if and only if condition (5) holds with $\mathrm{ind}(a) = 0$, i.e., when $p^{\gcd(2j,n)} - 1 \nmid \frac{p^n-1}{2}$. The latter holds if and only if $\frac{p^n-1}{p^{\gcd(2j,n)}-1}$ is odd and it remains to note that, by (8),

$$\frac{p^n-1}{p^{\gcd(2j,n)}-1} \equiv \frac{n}{\gcd(2j,n)} \pmod 2.$$ □

Assuming $j = n$ for an arbitrary n and $j = k$ for even $n = 2k$ in Theorem 1 leads directly to the Sidelnikov and Kasami p-ary bent functions. In the following two corollaries, we find the actual value of the Walsh transform coefficients for these two known classes of bent functions. This can be done applying directly (6) and Lemma 2, however, the corresponding character sums can also be computed directly and we demonstrate this in the proofs.

Corollary 2 (Sidelnikov). *For any nonzero $a \in \mathrm{GF}(p^n)$ and odd prime p, the function $f(x) = \mathrm{Tr}_n\left(ax^2\right)$ is a (weakly) regular bent function. Moreover, for $b \in \mathrm{GF}(p^n)$ the corresponding Walsh transform coefficient of $f(x)$ is equal to*

$$S_a(b) = \eta(a)(-1)^{n-1}p^{n/2}\omega^{-\mathrm{Tr}_n\left(\frac{b^2}{4a}\right)}, \quad if \ \ p \equiv 1 \pmod 4$$

and

$$S_a(b) = \eta(a)(-1)^{n-1}i^n p^{n/2}\omega^{-\mathrm{Tr}_n\left(\frac{b^2}{4a}\right)}, \quad if \ \ p \equiv 3 \pmod 4$$

where i is the complex primitive fourth root of unity and η is the quadratic character of $\mathrm{GF}(p^n)$.

Proof. From Theorem 1 it readily follows that Sidelnikov functions are (weakly) regular bent function for any nonzero $a \in \mathrm{GF}(p^n)$ (assume $j = n$). The exact value of the Walsh transform coefficients can be obtained using [7, Theorem 5.33] as following

$$S_a(b) = \sum_{x \in \mathrm{GF}(p^n)} \omega^{\mathrm{Tr}_n\left(ax^2-bx\right)} = \omega^{-\mathrm{Tr}_n\left(\frac{b^2}{4a}\right)}\eta(a)G(\eta, \chi_1)$$

where χ_1 is the canonical additive character of $\mathrm{GF}(p^n)$ and $G(\eta, \chi_1)$ is the Gaussian sum. By [7, Theorem 5.15],

$$G(\eta, \chi_1) = \begin{cases} (-1)^{n-1}p^{n/2}, & if \ \ p \equiv 1 \pmod 4 \\ (-1)^{n-1}i^n p^{n/2}, & if \ \ p \equiv 3 \pmod 4 . \end{cases}$$

In particular, when n is even

$$S_a(b) = -\eta(a)(-1)^{\frac{(p-1)n}{4}}p^{n/2}\omega^{-\mathrm{Tr}_n\left(\frac{b^2}{4a}\right)} = \pm p^{n/2}\omega^{-\mathrm{Tr}_n\left(\frac{b^2}{4a}\right)}$$

and, depending on a, p and n, $f(x)$ can be regular or weakly regular. Alternatively, when n is odd

$$p^{-n/2}\omega^{\mathrm{Tr}_n\left(\frac{b^2}{4a}\right)}S_a(b) = \begin{cases} \eta(a), & \text{if } p \equiv 1 \pmod 4 \\ \eta(a)(-1)^{(n-1)/2}i, & \text{if } p \equiv 3 \pmod 4 \end{cases}$$

and $f(x)$ can be regular or weakly regular in the first case and only weakly regular in the second.   □

Corollary 3 (p-ary Kasami). *Let $n = 2k$ and $a \in \mathrm{GF}(p^n)$ for an odd prime p. Then the function $f(x) = \mathrm{Tr}_n\left(ax^{p^k+1}\right)$ is bent if and only if $a + a^{p^k} \neq 0$. Moreover, if the latter condition holds then $f(x)$ is weakly regular and for $b \in \mathrm{GF}(p^n)$, the corresponding Walsh transform coefficient of $f(x)$ is equal to*

$$S_a(b) = -p^k\omega^{-\mathrm{Tr}_k\left(\frac{b^{p^k+1}}{a+a^{p^k}}\right)}.$$

Proof. It follows easily from Theorem 1 that $f(x)$ is bent. Indeed, for $n = 2k$ and $j = k$, the condition opposite to (5) is $p^n - 1 \mid \frac{p^n-1}{2} - \mathrm{ind}(a)(p^k - 1)$. On the other hand, any nonzero $a \in \mathrm{GF}(p^n)$ satisfies $a + a^{p^k} = 0$ (which is equivalent to $a^{p^k-1} = -1$) if and only if $\mathrm{ind}(a)(p^k - 1) \equiv \frac{p^n-1}{2} \pmod{(p^n - 1)}$.

The Walsh transform coefficient of the $f(x)$ evaluated at b is equal to

$$S_a(b) = \sum_{x\in\mathrm{GF}(p^n)} \omega^{\mathrm{Tr}_n\left(ax^{p^k+1}-bx\right)}$$

$$= \sum_{x\in\mathrm{GF}(p^n)} \omega^{\mathrm{Tr}_k\left(\left(a+a^{p^k}\right)x^{p^k+1}-bx-b^{p^k}x^{p^k}\right)}$$

$$= \sum_{x\in\mathrm{GF}(p^n)} \omega^{\mathrm{Tr}_k\left(a_1(x-\beta)^{p^k+1}-a_1\beta^{p^k+1}\right)}$$

$$\overset{(*)}{=} \omega^{-\mathrm{Tr}_k\left(a_1\beta^{p^k+1}\right)}\left((p^k+1)\sum_{z\in\mathrm{GF}(p^k)^*}\omega^{\mathrm{Tr}_k(a_1z)}+1\right)$$

$$= -p^k\omega^{-\mathrm{Tr}_k\left(\frac{b^{p^k+1}}{a+a^{p^k}}\right)}$$

where $a_1 = a + a^{p^k} \neq 0$ and $b = a_1\beta^{p^k}$ (thus, $\beta^{p^k+1} = \frac{b^{p^k+1}}{a_1^2}$) and $(*)$ holds since raising elements of $\mathrm{GF}(p^n)^*$ to the power of $p^k + 1$ is a $(p^k + 1)$-to-1 mapping onto $\mathrm{GF}(p^k)^*$ as proved in [12, Lemma 1]. Therefore, p-ary Kasami functions are weakly regular bent functions.   □

The class of bent functions due to Kumar and Moreno is an immediate consequence of Theorem 1 as well.

Corollary 4 (Kumar, Moreno [4]). *Let $n = ek$ for an odd integer e and take integer r in the range $1 \leq r \leq e$ with $\gcd(r, e) = 1$. Then the function*

$f(x) = \mathrm{Tr}_n\big(ax^{p^{rk}+1}\big)$ is a (weakly) regular bent function for any nonzero $a \in$ GF(p^n) and odd prime p.

Proof. If $n = ek$ and $j = rk$ then $\gcd(2j, n) = \gcd(2rk, ek) = k\gcd(2r, e) = k$. The condition opposite to (5) looks like $p^k - 1 \ \Big|\ \frac{p^{ek}-1}{2} - \mathrm{ind}(a)(p^{rk} - 1)$ that is equivalent to $p^k - 1 \ \Big|\ \frac{p^{ek}-1}{2}$ and holds if and only if $\frac{p^{ek}-1}{p^k-1}$ is even. However, $\frac{p^{ek}-1}{p^k-1} \equiv e \equiv 1 \pmod 2$, by (8) and since e is odd.   □

The following corollary contains an equivalent definition of the Kumar-Moreno class of p-ary bent functions that appeared in [13, Definition 7.5]. Note that if $\frac{n}{\gcd(j,n)}$ is odd (as required in Corollary 5) then $\frac{n}{\gcd(2j,n)}$ is odd as well and this is a condition of Corollary 1. Moreover, if one of the following equivalent conditions is fulfilled – either $\gcd(2j, n)$ divides j or $\frac{n}{\gcd(j,n)}$ is odd or $\gcd(2j, n) = \gcd(j, n)$ (in particular, this is true for an odd n) then (5) is equivalent to $p^{\gcd(j,n)} - 1 \ \Big|\ \frac{p^n-1}{2}$ which holds if and only if $\frac{p^n-1}{p^{\gcd(j,n)}-1}$ is odd. By (8), $\frac{p^n-1}{p^{\gcd(j,n)}-1} \equiv \frac{n}{\gcd(j,n)} \equiv 1 \pmod 2$. Thus, (5) holds for any nonzero $a \in$ GF(p^n) and $j \in \{1,\ldots,n\}$ meaning that all monomial quadratic bent functions in this case are of the Kumar-Moreno type. The Walsh transform coefficients for such functions were computed earlier in [6, Theorem 1 (i), Lemma 3.2].

Corollary 5. *Let j be an integer with $1 \le j \le n$ such that $\frac{n}{\gcd(j,n)}$ is odd. Then the function $f(x) = \mathrm{Tr}_n\big(ax^{p^j+1}\big)$ is a (weakly) regular bent function for any nonzero $a \in$ GF(p^n) and odd prime p. Moreover, for $b \in$ GF(p^n) the magnitude of $S_a(b)$ is equal to*

$$\omega^{\mathrm{Tr}_n\big(ax_0^{p^j+1}\big)} S_a(b) = \begin{cases} \eta(a)(-1)^{n-1}p^{n/2}, & \text{if } p \equiv 1 \pmod 4 \\ \eta(a)(-1)^{n-1}i^n p^{n/2}, & \text{if } p \equiv 3 \pmod 4 \end{cases}$$

where i is the complex primitive fourth root of unity, η is the quadratic character of GF(p^n) and x_0 is given by (7).

Proof. We will prove that the conditions of Corollaries 4 and 5 are equivalent. Denote $k = \gcd(j, n)$ and $e = n/k$ that is odd under the hypothesis. Since k divides j, let $j = rk$ with $\gcd(r, e) = 1$. Finally, the requirement $1 \le j \le n$ guarantees that $1 \le r \le e$.

Since $\frac{n}{\gcd(j,n)}$ is odd, it can be concluded that $v(n) \le v(j)$. Thus, we get in the conditions of Lemma 2 item (i) that gives us the magnitude of $S_a(b)$ and the rest follows from Theorem 1.   □

Note 1. Recall that conditions of Theorem 1 allow the values of j to be in the range $\{1,\ldots,n\}$. On the other hand, if $n = 2k+1$ or $n = 2k$ then for any $j > k$ we can write $(p^j+1)p^{n-j} = p^n+p^{n-j} \equiv p^{n-j}+1 \pmod{p^n-1}$. Thus, exponents $p^j + 1$ and $p^{n-j} + 1$ are cyclotomic equivalent and we can assume j to be in the range $\{0,\ldots,k\}$.

4 Conclusion

We considered quadratic functions over $GF(p^n)$ (p is odd) that can be represented in the univariate form as $f(x) = \mathrm{Tr}_n(ax^{p^j+1})$ with $a \in GF(p^n)$. We proved the criterion on a for such a function to be generalized bent. All the previously known monomial quadratic bent functions appear as a particular case of our general result. Moreover, we obtained the explicit expression for the Walsh transform coefficients of a bent function in this class. In particular, this determines the dual of a monomial quadratic bent function which is also a bent function. This dual function belongs to a new, previously unknown, class of generalized bent functions. A challenging open problem is to find the Walsh transform coefficients (the dual function, in particular) for a quadratic bent function consisting of more than a single term in its univariate representation. By (3), this task requires finding x_0 with $F^*(x_0) = -b^{p^l}$ which seems to be difficult in general.

References

1. Kumar, P.V., Scholtz, R.A., Welch, L.R.: Generalized bent functions and their properties. Journal of Combinatorial Theory, Series A 40(1), 90–107 (1985)
2. Hou, X.D.: p-Ary and q-ary versions of certain results about bent functions and resilient functions. Finite Fields and Their Applications 10(4), 566–582 (2004)
3. Helleseth, T., Kholosha, A.: Monomial and quadratic bent functions over the finite fields of odd characteristic. IEEE Trans. Inf. Theory 52(5), 2018–2032 (2006)
4. Kumar, P.V., Moreno, O.: Prime-phase sequences with periodic correlation properties better than binary sequences. IEEE Trans. Inf. Theory 37(3), 603–616 (1991)
5. Liu, S.C., Komo, J.J.: Nonbinary Kasami sequences over $GF(p)$. IEEE Trans. Inf. Theory 38(4), 1409–1412 (1992)
6. Coulter, R.S.: Further evaluations of Weil sums. Acta Arithmetica 86(3), 217–226 (1998)
7. Lidl, R., Niederreiter, H.: Finite Fields. Encyclopedia of Mathematics and its Applications, vol. 20. Cambridge University Press, Cambridge (1997)
8. Coulter, R.S.: Explicit evaluations of some Weil sums. Acta Arithmetica 83(3), 241–251 (1998)
9. Draper, S., Hou, X.D.: Explicit evaluation of certain exponential sums of quadratic functions over $\mathbb{F}_{p^n}$, p odd. arXiv:0708.3619v1 (2007)
10. Helleseth, T.: Some results about the cross-correlation function between two maximal linear sequences. Discrete Mathematics 16(3), 209–232 (1976)
11. Baumert, L., McEliece, R.J.: Weights of irreducible cyclic codes. Information and Control 20(2), 158–175 (1972)
12. Delsarte, P., Goethals, J.M.: Tri-weight codes and generalized Hadamard matrices. Information and Control 15(2), 196–206 (1969)
13. Helleseth, T., Kumar, P.V.: Sequences with low correlation. In: Pless, V., Huffman, W. (eds.) Handbook of Coding Theory, vol. 2, pp. 1765–1853. Elsevier, Amsterdam (1998)

A New Family of Gold-Like Sequences

Xiaohu Tang[1], Tor Helleseth[2], Lei Hu[3], and Wenfeng Jiang[3,*]

[1] The Provincial Key Lab of Information Coding and Transmission, Institute of
Mobile Communications, Southwest Jiaotong University, Chengdu, China
xhutang@ieee.org
[2] The Selmer Center, Department of Informatics, University of Bergen PB 7803,
N-5020 Bergen, Norway
tor.helleseth@ii.uib.no
[3] The State Key Laboratory of Information Security, Graduate School of Chinese
Academy of Sciences, Beijing 100049, China
{hu,wfjiang}@is.ac.cn

Abstract. In this paper, for a positive odd integer n, a new family of
binary sequences with $2^n + 1$ sequences of length $2^n - 1$ taking three level
nontrivial correlations -1 and $-1 \pm 2^{(n+1)/2}$ is presented, whose correlation distribution is the same as that of the well-known Gold sequences.
This family may be considered as a new class of Gold-like sequences.

Keywords: Binary sequence, Gold sequence, Gold-like sequence.

1 Introduction

Sequence sets with good correlations are widely used in many applications, for
example Code Division Multiple Access (CDMA) communication systems and
cryptography system [2]. Since the late sixties, many families of binary sequences
of length $2^n - 1$ have been found [4], where n is a positive integer. Among them,
when n is odd, the well-known Gold sequence family is the oldest binary family
of $2^n + 1$ sequences having three level out of phase auto- and cross-correlation
(nontrivial correlation) values -1 and $-1 \pm 2^{(n+1)/2}$ [3]. The family is optimal
with respect to the Sidelnikov bound [7]. In 1990s, Boztas and Kumar discovered
an optimal sequence family [1], which has the same correlation distribution as
that of Gold sequences but larger linear span, and therefore named it Gold-like
sequences. Later, Kim and No further generalized the Gold-like sequences to
GKW-like sequences by the quadratic form technique [5].

In this paper, we use quadratic form technique to generalize Gold-like sequences. As a result, we get a new family of optimal binary sequences with
$2^n + 1$ sequences of length $2^n - 1$, whose correlation distribution is identical to

* This work of X.H. Tang was supported by the Program for New Century Excellent
Talents in University (NCET) under Grant 04-0888, and the Foundation for the
Author of National Excellent Doctoral Dissertation of PR China (FANEDD) under
the Grant 200341. The research of T. Helleseth was supported by the Norwegian
Research Council.

S.W. Golomb et al. (Eds.): SSC 2007, LNCS 4893, pp. 62–69, 2007.
© Springer-Verlag Berlin Heidelberg 2007

that of Gold sequence for n odd. In this sense, this family can be seen as a new class of Gold-like sequences.

The paper is organized as follows. In Section 2, we give the notation and the necessary preliminaries required for the subsequent sections. In Section 3, we present the new family of binary sequences and our main result. In Section 4, we investigate two quadratic forms involved in computing the correlation function of the new sequences. Finally in Section 5, we give the proof of our main result.

2 Preliminaries

For convenience, the following notations are used throughout this paper

- $\mathbf{F}_q$ is the finite fields with q elements, and $\mathbf{F}_q^*$ denotes its multiplicative group.
- n, m, and e are odd integers with $n = em$ and $m \geq 3$.
- ζ is an element in $\mathbf{F}_{2^e}$ and $\zeta \neq 1$.
- $\{\zeta_0, \zeta_1, \cdots, \zeta_{2^n-1}\}$ is an enumeration of the elements in $\mathbf{F}_{2^n}$.
- $tr_e^n(x) = \sum_{l=0}^{m-1} x^{2^{el}}$ and $tr_1^n(x) = \sum_{l=0}^{n-1} x^{2^l}$ are the trace functions from $\mathbf{F}_{2^n}$ to $\mathbf{F}_{2^e}$ and $\mathbf{F}_{2^n}$ to $\mathbf{F}_2$ respectively.

Let $a = (a(0), a(1), \cdots, a(N-1))$ and $b = (b(0), b(1), \cdots, b(N-1))$ be two binary sequences of period N, we define the periodic correlation between a and b as

$$R_{a,b}(\tau) = \sum_{t=0}^{N-1} (-1)^{a(t)+b(t+\tau)}, \quad 0 \leq \tau < N.$$

Basically, in this paper the correlation functions of all the sequences could be transformed into the following form:

$$F(\lambda) = \sum_{x \in \mathbf{F}_{2^n}} (-1)^{f(x)+tr_1^n(\lambda x)}, \tag{1}$$

where $f(x)$ is a quadratic form in $\mathbf{F}_{2^n}$ over $\mathbf{F}_2$ and $\lambda \in \mathbf{F}_{2^n}$. Usually, $F(\lambda)$ is also called the *trace transform* of $f(x)$.

Definition 1. *Let* $x = \sum_{i=1}^n x_i \alpha_i$ *where* $x_i \in \mathbf{F}_2$ *and* α_i, $i = 1, 2, \ldots, n$, *is a basis for* $\mathbf{F}_{2^n}$ *over* $\mathbf{F}_2$. *Then the function* $f(x)$ *over* $\mathbf{F}_{2^n}$ *to* $\mathbf{F}_2$ *is a quadratic form if it can be expressed as*

$$f(x) = f(\sum_{i=1}^n x_i \alpha_i) = \sum_{i=1}^n \sum_{j=1}^n b_{i,j} x_i x_j,$$

where $b_{i,j} \in \mathbf{F}_2$, *that is* $f(x)$ *is a homogeneous polynomial of degree 2 on* $\mathbf{F}_2^n$.

The quadratic form has been well analyzed in [6]. It should be noted that a quadratic form is completely determined by its rank, which is defined as the minimum number of variables required to represent the function under the non-singular coordinate transformations.

The following lemma is useful to establish the relationship between the trace transform and the rank of a quadratic form.

Lemma 1 ([4]). *Let $f(x)$ be a quadratic form in $\mathbf{F}_{2^n}$ over $\mathbf{F}_2$. If the rank of $f(x)$ is $2r$, $2 \leq 2r < n$, then the distribution of the trace transform values is given by*

$$F(\lambda) = \begin{cases} 2^{n-r}, & 2^{2r-1} + 2^{r-1} \ times \\ 0, & 2^n - 2^{2r} \ times \\ -2^{n-r}, & 2^{2r-1} - 2^{r-1} \ times \end{cases} \tag{2}$$

From Lemma 1, the rank of the quadratic form is crucial to compute its trace transform. In [6], it is shown that the rank $2r$ is related to the number of solution of $x \in \mathbf{F}_{2^n}$ to

$$B(x, z) = f(x) + f(z) + f(x + z) = 0, \forall z \in \mathbf{F}_{2^n}. \tag{3}$$

More precisely, suppose that the number of solutions is $\mathcal{N}$, then $2r = n - \log_2 \mathcal{N}$. In terminology, $B(x, z) = f(x) + f(z) + f(x + z)$ is called *symplectic form* of quadratic form $f(x)$.

Bozats and Kumar [1] studied the quadratic form $p(x)$ defined as

$$p(x) = \sum_{l=1}^{\frac{n-1}{2}} tr_1^n (x^{2^l+1}).$$

Lemma 2 ([1]). *The associated symplectic form of $p(x)$ is*

$$B(x, z) = p(x) + p(z) + p(x + z) = tr_1^n [z(tr_1^n(x) + x)]. \tag{4}$$

Based on $p(x)$, Bozats and Kumar obtained a family of binary sequence.

Definition 2 ([1]). *Sequences family $\mathcal{G} = \{g_i, i = 0, 1, \cdots, 2^n\}$ of length $2^n - 1$ is defined by*

$$g_i(t) = \begin{cases} tr_1^n (\zeta_i \alpha^t) + p(\alpha^t), & 0 \leq i < 2^n \\ tr_1^n (\alpha^t), & i = 2^n \end{cases} \tag{5}$$

Theorem 3 ([1]). *The correlation distribution of family $\mathcal{G}$ is*

$$R_{i,j}(\tau) = \begin{cases} -1 + 2^n, & 2^n + 1 \ times \\ -1, & 2^{3n-1} + 2^{2n} - 2^n - 2 \ times \\ -1 + 2^{\frac{n+1}{2}}, & (2^{2n} - 2)(2^{n-2} + 2^{\frac{n-3}{2}}) \ times \\ -1 - 2^{\frac{n+1}{2}}, & (2^{2n} - 2)(2^{n-2} - 2^{\frac{n-3}{2}}) \ times \end{cases}$$

This family is called Gold-like sequences since it possesses the same correlations as those of Gold sequences [3]. In 2003, Kim and No generalized the quadratic form $p(x)$ to

$$q(x) = \sum_{l=1}^{\frac{m-1}{2}} tr_1^n(x^{2^{el}+1}),$$

whose symplectic form is the following.

Lemma 4 (Kim and No [5]). *The associated symplectic form of $q(x)$ is*

$$B(x,z) = q(x) + q(z) + q(x+z) = tr_1^n[z(tr_e^n(x) + x)]. \tag{6}$$

3   Main Result

In this section, we construct a family of sequences based on two quadratic forms $p(x)$ and $q(\zeta x)$ as follows.

Definition 3. *The binary family $\mathcal{U}$ of sequences $\{u_i, i = 0, 1, \cdots, 2^n\}$ of length $2^n - 1$ is defined by*

$$u_i(t) = \begin{cases} tr_1^n(\zeta_i \alpha^t) + p(\alpha^t) + q(\zeta \alpha^t), & 0 \le i < 2^n \\ tr_1^n(\alpha^t), & i = 2^n \end{cases}. \tag{7}$$

For the correlation property of the family $\mathcal{U}$, we have the main result below.

Theorem 5. *Family $\mathcal{U}$ has the following properties:*

1. *The maximal absolute value of nontrival correlation of Family $\mathcal{U}$ is bounded by $R_{\max} \le 1 + 2^{\frac{n+1}{2}}$. The family is optimal with respect to the Sidelnikov bound.*
2. *The correlation distribution of Family $\mathcal{U}$ is as follows:*

$$R_{i,j}(\tau) = \begin{cases} -1 + 2^n, & 2^n + 1 \ times \\ -1, & 2^{3n-1} + 2^{2n} - 2^n - 2 \ times \\ -1 + 2^{\frac{n+1}{2}}, & (2^{2n} - 2)(2^{n-2} + 2^{\frac{n-3}{2}}) \ times \\ -1 - 2^{\frac{n+1}{2}}, & (2^{2n} - 2)(2^{n-2} - 2^{\frac{n-3}{2}}) \ times \end{cases}.$$

3. *The maximal linear complexity of Family $\mathcal{U}$ is bounded by $n(n + 1)/2$.*

Remark 1. *When $\zeta = 0$, the sequences Family $\mathcal{U}$ in Definition 3 is Gold-like family in (5). When $\zeta \in \mathbf{F}_{2^e} \setminus \{0, 1\}$, the new sequences Family $\mathcal{U}$ in Definition 3 has the same correlation distribution as those of Gold sequences and also Gold-like sequences.*

4   Two Quadratic Forms and Their Trace Transforms

A. Quadratic form $p(x) + q(\zeta x)$

From (4) and (6), the symplectic form of $p(x) + q(\zeta x)$ is

$$B(x, z) = tr_1^n[z(tr_1^n(x) + \zeta tr_e^n(\zeta x) + x + \zeta^2 x)].$$

According to (3), for computing the rank of the quadratic form $p(x) + q(\zeta x)$, it suffices to find the number of solutions to

$$tr_1^n(x) + \zeta tr_e^n(\zeta x) + x + \zeta^2 x = 0.$$

Let $tr_e^n(x) = a$. Then

$$x = \frac{tr_1^e(a) + \zeta^2 a}{1 + \zeta^2},$$

which indicates $x \in \mathbf{F}_2^e$. Plugging it into $tr_e^n(x) = a$, we have

$$a = \frac{tr_1^e(a) + \zeta^2 a}{1 + \zeta^2}.$$

It follows that $a = tr_1^e(a)$, which has two solutions $a = 0$ or 1. That is, the rank of quadratic form $p(x) + q(\zeta x)$ is $2r = n - 1$.

B. Quadratic form $p(x) + q(\zeta x) + p(\delta x) + q(\zeta \delta x)$

Let $\delta \neq 1 \in \mathbf{F}_{2^n}$ be a constant, we study the quadratic form $p(x) + q(\zeta x) + p(\delta x) + q(\zeta \delta x)$. For this quadratic form, by (4) and (6) the associated symplectic form is

$$B(x, z) = tr_1^n[z(\delta tr_1^n(\delta x) + tr_1^n(x) + \zeta \delta tr_e^n(\zeta \delta x) + \zeta tr_e^n(\zeta x) + (1 + \zeta^2)(1 + \delta^2)x)].$$

Similarly, we need to count the solutions to

$$\delta tr_1^n(\delta x) + tr_1^n(x) + \zeta \delta tr_e^n(\zeta \delta x) + \zeta tr_e^n(\zeta x) + (1 + \zeta^2)(1 + \delta^2)x = 0.$$

Let $tr_e^n(x) = a$ and $tr_e^n(\delta x) = b$. Then

$$x = \frac{\delta tr_1^e(b) + tr_1^e(a) + \zeta^2 \delta b + \zeta^2 a}{(1 + \zeta^2)(1 + \delta^2)}$$

$$= \frac{\zeta^2 a + tr_1^e(a) + \delta(\zeta^2 b + tr_1^e(b))}{(1 + \zeta^2)(1 + \delta^2)}. \tag{8}$$

Let $X = tr_e^n(\frac{1}{1+\delta})$. Plugging (8) into $tr_e^n(x) = a$ and $tr_e^n(\delta x) = b$, we have

$$(\zeta^2 a + tr_1^e(a) + \zeta^2 b + tr_1^e(b))X^2 + (\zeta^2 b + tr_1^e(b))X = a(\zeta^2 + 1), \tag{9}$$

and

$$(\zeta^2 a + tr_1^e(a) + \zeta^2 b + tr_1^e(b))X^2 + (\zeta^2 a + tr_1^e(a))X = tr_1^e(b) + b. \tag{10}$$

There are three cases.

1. If $X = 0$, then $a = 0$ and $tr_1^e(b) = b$. Obviously, there are two solutions $(a, b) = (0, 0)$ and $(0, 1)$;
2. If $X = 1$, then $tr_1^e(a) = a$ and $b = 0$. It is easy to see that there are two solutions $(a, b) = (0, 0)$ and $(1, 0)$;
3. If $X = c \in \mathbf{F}_{2^e} \setminus \{0, 1\}$, then

$$\zeta^2 a + tr_1^e(a) + \zeta^2 b + tr_1^e(b) = \frac{a(\zeta^2 + 1) + b + tr_1^e(b)}{c}. \tag{11}$$

Replacing $a + tr_1^e(a) + b + tr_1^e(b)$ in (9) and (10) with the right-hand side of (11), we get two equations, i.e.,

$$c(a + b) = a,$$
$$c(a + b + tr_1^e(a) + tr_1^e(b)) = b + tr_1^e(b).$$

We discuss them in four possibilities:

3.1 $tr_1^e(a) = 0$ and $tr_1^e(b) = 0$. Then

$$c(a + b) = a,$$
$$c(a + b) = b.$$

Immediately, $a = b = 0$.

3.2 $tr_1^e(a) = 1$ and $tr_1^e(b) = 1$. Then

$$c(a + b) = a,$$
$$c(a + b) = b + 1.$$

It follows that $a = c$ and $b = c + 1$, which lead a contradiction with $tr_1^e(a) = tr_1^e(b) = 1$.

3.3 $tr_1^e(a) = 1$ and $tr_1^e(b) = 0$. Then

$$c(a + b) = a,$$
$$c(a + b + 1) = b.$$

We have $a = c^2$, $b = c^2 + c$, and $tr_1^e(c) = 1$.

3.4 $tr_1^e(a) = 0$ and $tr_1^e(b) = 1$. Then

$$c(a + b) = a,$$
$$c(a + b + 1) = b + 1.$$

Immediately, $a = c^2 + c$, $b = c^2 + 1$, and $tr_1^e(c) = 0$.

Thus, for $X = c \in \mathbf{F}_{2^e} \setminus \{0, 1\}$, the associated symplectic form $B(x, z)$ has

1. Two solutions $(a, b) = (0, 0)$ and $(c^2, c^2 + c)$ when $tr_1^e(c) = 1$;
2. Two solutions $(a, b) = (0, 0)$ and $(c^2 + c, c^2 + 1)$ when $tr_1^e(c) = 0$.

In summary, the rank of the quadratic form $p(x) + q(\zeta x) + p(\delta x) + q(\zeta \delta x)$ is therefore $2r = n - 1$.

5 Proof of Theorem 5

We investigate the correlation function $R_{i,j}(\tau)$ between u_i and u_j in five cases.

Case 1. $0 \leq i = j \leq 2^n$ and $\tau = 0$:
  In this trivial case, $R_{i,i}(0) = 2^n - 1$.

Case 2. $i = j = 2^n$ and $0 < \tau < 2^n - 1$:
  Since u_{2^n} is an m-sequence, we have $R_{i,j}(\tau) = -1$.

Case 3. $0 \leq i \neq j < 2^n$ and $\tau = 0$:
  In this case, $u_i(t) + u_j(t) = tr_1^n((\zeta_i + \zeta_j)\alpha^t)$, and therefore, $R_{i,j}(0) = -1$ again from the auto-correlation property of m-sequence $(tr_1^n(\alpha^t), t = 0, 1, \cdots, 2^n - 2)$.

Case 4. $0 \leq i < 2^n$ and $j = 2^n$ (or $(i = 2^n$ and $0 \leq j < 2^n))$:
  For a fixed $0 \leq \tau < 2^n - 1$,

$$R_{i,2^n}(\tau) = \sum_{x \in \mathbf{F}_2^n} (-1)^{tr_1^n((\zeta_i + \delta)x) + p(x) + q(\zeta x)} - 1,$$

where $\delta = \alpha^\tau$.
  In Section 4, we proved that the rank of the quadratic form $p(x) + q(\zeta x)$ is $n - 1$. Consequently, it follows from Lemma 1 that the distribution of the correlations for a fixed $0 \leq \tau < 2^n - 1$ is

$$R_{i,2^n}(\tau) = \begin{cases} -1 + 2^{\frac{n+1}{2}}, & 2^{n-2} + 2^{\frac{n-3}{2}} \ times \\ -1, & 2^n - 2^{n-1} \ times \\ -1 - 2^{\frac{n+1}{2}}, & 2^{n-2} - 2^{\frac{n-3}{2}} \ times \end{cases}$$

As τ varies over the range $0 \leq \tau < 2^n - 1$, the distribution of the correlations becomes

$$R_{i,2^n}(\tau) = \begin{cases} -1 + 2^{\frac{n+1}{2}}, & (2^{n-2} + 2^{\frac{n-3}{2}})(2^n - 1) \ times \\ -1, & (2^n - 2^{n-1})(2^n - 1) \ times \\ -1 - 2^{\frac{n+1}{2}}, & (2^{n-2} - 2^{\frac{n-3}{2}})(2^n - 1) \ times \end{cases}$$

The same distribution holds for $R_{2^n,j}(\tau)$, the case of $i = 2^n$ and $0 \leq j < 2^n$.

Case 5. $0 \leq i, j < 2^n$ and $0 < \tau < 2^n - 1$:
  For a fixed $0 \leq \tau < 2^n - 1$, let $\delta = \alpha^\tau$. Then, the correlation function is

$$R_{i,j}(\tau) = \sum_{x \in \mathbf{F}_{2^n}} (-1)^{tr_1^n((\zeta_i + \zeta_j \delta)x) + p(x) + q(\zeta x) + p(\delta x) + q(\zeta \delta x)} - 1.$$

In Section 4, it is shown that the rank of quadratic form $p(x) + q(\zeta x) + p(\delta x) + q(\zeta \delta x)$ is $n - 1$. Similar to the distribution in Case 4, the correlation distribution can be computed from Lemma 1 as

$$R_{i,j}(\tau) = \begin{cases} -1 + 2^{\frac{n+1}{2}}, & (2^{n-2} + 2^{\frac{n-3}{2}})2^n(2^n - 2) \ times \\ -1, & (2^n - 2^{n-1})2^n(2^n - 2) \ times \\ -1 - 2^{\frac{n+1}{2}}, & (2^{n-2} - 2^{\frac{n-3}{2}})2^n(2^n - 2) \ times \end{cases}$$

where τ ranges over $0 < \tau < 2^n - 1$ and i, j vary from 0 to $2^n - 1$, respectively.

Collecting the results of the above five cases, we obtain the distribution of the correlation values of the sequences Family $\mathcal{U}$.

References

1. Boztas, S., Kumar, P.V.: Binary sequences with Gold-like correlation but larger linear span. IEEE Trans. Inform. Theory 40, 532–537 (1994)
2. Fan, P.Z., Darnell, M.: Sequence Design for Communications Applications. John Wiley, Chichester (1996)
3. Gold, R.: Maximal recursive sequences with 3-valued recursive crosscorrelation functions. IEEE Trans. on Inform. Theory 14, 154–156 (1968)
4. Helleseth, T., Kumar, P.V.: Sequences with low correlation. In: Pless, V., Huffman, C. (eds.) Handbook of Coding Theory, Elsevier, Amsterdam, The Netherlands (1998)
5. Kim, S.H., No, J.S.: New families of binary sequences with low crosscorrelation property. IEEE Trans. Inform. Theory 49, 3059–3065 (2003)
6. Lidl, R., Niederreiter, H.: Finite Fields. In: Encyclopedia of Mathematics, vol. 20, Cambridge University Press, Cambridge (1983)
7. Sidelnikov, V.M.: On mutual correlation of sequences. Soviet Math. Dokl. 12, 197–201 (1971)

Sequencings and Directed Graphs with Applications to Cryptography

Lothrop Mittenthal

Custom Cryptography Company
240 Los Padres Drive
Thousand Oaks, CA 91361

Abstract. A motivation for this paper is to find a practical mechanism for generating permutations of a finite set of consecutive positive integers so that the resultant spacings between originally consecutive numbers i and $i + i$, are now different for each i. This is equivalent to finding complete Latin squares. The ordered set of spacings in such a permutation is called a sequencing. The set of partial sums of the terms in a sequencing is called a directed terrace. For lack of standard terminology, the associated permutations here are called quick trickles.

This paper concerns methods of finding such sequencings, in part by finding constraints on their existence, so that search time can be substantially reduced. A second approach is to represent a quick trickle permutation as a directed graph. The sequencings then are represented by chords of different lengths. Various methods can be used to rearrange the chords and obtain additional sequencings and groups of quick trickle permutations.

It is well known that complete Latin squares of size $n \times n$ can be found if n is even but not if n is odd. This is equivalent to saying that the group of integers $(1, 2, 3, \ldots, n)$ under the operation of multiplication mod n is sequenceable if and only if n is even. A more general concept is introduced which is termed quasi-sequenceable. This applies to both even and odd sizes.

The application to block encryption, or the so-called substitution/ permutation system is briefly described. The net result is interround mixing, deterministically generated under key control, and quickly replaceable with a new pattern of equal merit.

Currently, typical block encryption systems use algorithms which are fixed and publicly known. The motivation here is to develop block encryption systems using algorithms which are variable, secret, generated and periodically changed by the key.

Keywords: Block encryption, Complete Latin square, Directed graph, Directed terrace, Inter-round mixing, Quick trickle permutation, Sequenceable group, Sequencing, S/P network.

1 Introduction

A motivation for this paper is to find a practical mechanism for generating permutations of a finite set of consecutive positive integers so that the resultant

S.W. Golomb et al. (Eds.): SSC 2007, LNCS 4893, pp. 70–81, 2007.

spacings between originally consecutive numbers i and $i + 1$, are now different for each i. This is equivalent to finding complete Latin squares, that is, $n \times n$ arrays of numbers in which there are n distinct entries in each row and column, arranged such that the order or permutation in each row is different as is the order in each column. There is considerable literature on Latin squares, including orthogonal Latin squares with transversals plus generalizations and variations, such as, Tuscan, Roman, Florentine and Vatican squares. (See references 4. and 5.) However, this paper is primarily concerned with complete Latin squares because of their applications to cryptography, specifically to the process of interround mixing in block substitution systems.

Typically, a block substitution system consists of alternating steps of encrypting conveniently sized sub-blocks in individual substitution boxes (S-boxes) and permuting the partly encrypted data. It is usually impractical to encrypt large blocks in one step, and the individual S-boxes will, normally, have all different tables. In some systems, the full set of individual bits are permuted after each round; however, it is simpler and quicker to permute the partially encrypted sub-blocks. For good mixing, each sub-block and its partially encrypted successor should pass through each S-box (Fig. 1).

The above example is for four rounds with four sub-blocks, 4×4 Latin squares, but the pattern holds for any even number. The complete Latin square in Fig. 2 gives the most thorough mixing of the entries in the Latin square. This mixing pattern could be specified by a table, as above, but this can be quite cumbersome for large numbers of sub-blocks and rounds. A more efficient mechanism is to

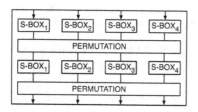

Fig. 1. Typical S/P Networks for Block Encryption

A	B	C	D	ROUNDS 1
D	A	B	C	2
C	D	A	B	3
B	C	D	A	4

Latin square - same neighbors, same successors

A	B	C	D	ROUNDS 1
C	A	D	B	2
B	D	A	C	3
D	C	B	A	4

Complete Latin square - different neighbors, different successors

Fig. 2. Types of Latin squares

find permutations which will directly generate these mixing patterns in the S/P networks.

2   Quick Trickle Permutations

The complete Latin square in Fig. 2 can be derived by permuting the entries in the first row, ABCD, by selected powers of a permutation in which the spacings between integers which were adjacent in their natural order, are now all different. Consider the permutation $g = (1243)$. The spacings are from 1 to 2 $a_1 = 1$, from 2 to 3 $a_2 = 2$ and from 3 to 4 $a_3 = 3$. If we now apply cumulative powers of the permutation $g = (1243)$ to the first row ABCD, we obtain:

$$
\begin{array}{lll}
g^0 = (1)(2)(3)(4) & \text{ABCD} & \text{row 1 or round 1} \\
g^1 = (1\ 2\ 4\ 3) & \text{CADB} & \text{row 2 or round 2} \\
g^2 \circ g^1 = g^3 = (1\ 3\ 4\ 2) & \text{BDAC} & \text{row 3 or round 3} \\
g^3 \circ g^2 \circ g^1 = g^2 = (1\ 4)(2\ 3) & \text{DCBA} & \text{row 4 or round 4}
\end{array}
$$

where the exponents are added modulo 4 and where the symbol "∘" means composition of the permutations. Note that the sequence of powers applied to g is $0, 1, 3, 2$ applied to the original S-box arrangement. This will be generalized in successive sections. Further, because there does not appear to be any standard name for permutations with distinct spacings, in this paper we will use the following:

Definition 1. *A quick trickle permutation is one in which the spacings a_i from the location of i to the location of $i+1$ are all different.*

More generally, consider an S/P block substitution system of $n = 2s$ sub-blocks and the same number of rounds. By definition, a quick trickle permutation will be characterized by a set of $n - 1$ spacings, $a_1, \cdots, a_{n-1}$ where all are different. For convenience, we include the trivial spacings of 1 to itself in the permutation, that is, $a_0 \equiv 0$. The set of integers $\{a_0, a_1, a_2, \cdots, a_{n-1}\}$, is, thus, an ordered arrangement or permutation of the n integers $\{0, 1, 2, \cdots, n - 1\}$, that is, the integers modulo n. This is the group Z_n with the group operation of addition modulo n. We can also define

$$
b_j = \sum_{i=0}^{j} a_i \pmod{n}.
$$

Clearly, b_j is the net spacing in the quick trickle permutation from the initial integer 1 to the integer $j + 1$. This set of partial sums of integers mod n, in some arbitrary order, will, generally, not be distinct; however, if the a_i are defined by a quick trickle permutation, the b_i must be distinct, otherwise, two or more numbers would occupy the same position relative to 1. Also for $n \equiv 2s$, since the order of addition is immaterial:

$$b_{n-1} = \sum_{i=0}^{n-1} a^i = \sum_{i=0}^{n-1} i = \frac{n(n-1)}{2} = \frac{n(n-2)}{2} + \frac{n}{2} \equiv s \pmod{n} \qquad (1)$$

This proves the following:

Proposition 1. *In a quick trickle permutation of the integers* $(1, 2, \cdots, n = 2s)$, *the net spacing from 1 to n is s.*

If g is a quick trickle permutation and k is an integer where $GCD(k, n) = 1$, then g^k is also a quick trickle permutation. If not, there would be a pair of spacings $a_i \neq a_j$ such that $ka_i \equiv ka_j \pmod{n}$, or $k(a_i - a_j) = cn$ for some positive integer c. Then, k divides c and $a_i - a_j = \frac{c}{k}n > n$ which is not possible. Thus, any quick trickle permutation defines a group of permutations in which those without subcycles are also quick trickle permutations. For those powers yielding permutations with cycles, all spacings a_i are not defined, but those which are defined, are all different.

3   Sequenceable Groups

The so-called quick trickle permutations are related to the concept of a sequenceable group. See reference 3.

Definition 2. *A finite group* $(G, \circ)$ *with group elements* $a_0 = e$ *(the identity),* $a_1, a_2, \cdots, a_{n-1}$ *and group operation "$\circ$" is called sequenceable if the group elements can be arranged in such an order that the partial sums (products)* $b_0 = a_0, b_1 = a_0 \circ a_1, b_2 = a_0 \circ a_1 \circ a_2$, *etc., are all different. The ordered set* $a_0, a_1, a_2, \cdots$ *is called a sequencing, and the set of partial sums* $b_0, b_1, b_2, \cdots$ *is called a directed terrace.*

There is a fair amount of theory on the existence of sequenceable groups, but for our purposes it suffices to know that the groups of integers modulo n, Z_n, for n even and addition modulo n as the group operation, are sequenceable. In the case of quick trickle permutations, the ordered set of distinct spacings $\{a_i\}$ is a sequencing. The fact that they are derived from a permutation guarantees that the partial sums are a directed terrace.

4   Applications to Inter-Round Mixing

To apply this to block encryption where one typically uses an even number of different block encryption tables, one arranges the tables in initial order and in successive rounds permutes the order of the tables with a quick trickle permutation raised to successive powers where the exponents are a directed terrace. The directed terrace need not be derived from the same quick trickle permutation (QTP). These two sequences of integers can be part of the encryption key and changed at the end of each code validity interval. In summary, the $2n$ integers added to the encryption key will consist of n integers of a QTP and n integers of a directed terrace.

5   Quasi-Complete Latin Squares and Quasi-Sequencings

Complete Latin squares do not exist for all values of n. As shown in ref. 3, a sufficient condition for the existence of a complete Latin square is that there exists a sequenceable group of order n. Commutative (Abelian) sequenceable groups of odd order do not exist. A substitute for the complete Latin square is the quasi-complete Latin square.

Definition 3. Ref. 3: *An $n \times n$ Latin square is said to be quasi-row-complete if the pairs of adjacent elements which occur in the rows include each unordered pair of distinct elements exactly twice. It is called quasi-column-complete if the adjacent (succeeding or preceding) pairs of elements in the columns include each unordered pair of distinct elements exactly twice. A Latin square is called quasi-complete if it is both quasi-row-complete and quasi-column-complete.*

6   Searching for Sequencings

Since any sequencing defines a quick trickle permutation and vice versa, one approach would be to examine arbitrary permutations and reject those not qualifying. This becomes a formidable task. There are $n!$ permutations of the integers $1, 2, \cdots, n$; however, without loss of generality, one can specify the integer 1 in the leftmost position, leaving $(n-1)!$ possibilities. For a quick trickle permutation, the integer n always appears $s = \frac{n}{2}$ positions to the right of 1, leaving $(n-2)!$ possibilities. The same is true for the directed terrace since the first number is 0 and the last number is $b_{n-1} = \frac{n}{2} = s$. There are other restrictions that reduce the number of possibilities. For example, in the sequencing $\{a_0 = 0, a_1, a_2, \cdots, a_{n-1}\}$, there can be no subset of consecutive numbers which sum to 0 mod n, which is the same as saying that no such subset is an unequal partition of $n, 2n, \cdots, (s-1)n$. If there were such a partition, two of the partial sums b_i would be the same. Similarly, no partial sum $b_i = s$ mod n is possible since s would not be the last term in the directed terrace. These rules are helpful in the sense of eliminating possible candidates for quick trickle permutations but not in constructing them; however, one such sequence that is known to be a directed terrace for any $n = 2s$ is:

$$0, n-1, 1, n-2, 2, n-3, 3, \cdots$$

See reference 1. The corresponding sequencing is:

$$0, n-1, 2, n-3, 4, n-5, 6, \cdots$$

and the corresponding quick trickle permutation is:

$$g = (1, 3, 5, \cdots, n-1, n, n-2, 2)$$

Another simple example of a quick trickle permutation is:

$$p = (1, 2, 4, 6, \cdots, n, n-1, n-3, 3) \tag{2}$$

$a_i = i$ for i odd and $a_i = n - i$ for i even. For $n = 2s$, the a_i are all different, and since they are derived from a permutation, the $b_i = \sum_{j=0}^{i} a_j$ must all be distinct. Also, $g = p^{n-1} = (1, 3, 5, \cdots, n - 1, n, n - 2, \cdots, 2)$.

As pointed out above, a prospective permutation of the integers $0, 1, 2, \cdots, n - 1$ with addition modulo $n = 2s$ cannot be a sequencing unless no subset of consecutive numbers is an unequal partition of a multiple of n, less than or equal to $(s - 1)n$. It is tempting to examine as a candidate for spacings $a_i = i, 0 \leq i \leq n - 1$, that is, the possibility that $\{0, 1, 2, 3, \cdots, n - 1\}$ is a sequencing under addition modulo n.

Proposition 2. For $n = 2s$, the sequence $\{0, 1, 2, \cdots, n - 1\}$ is a sequencing if and only if $n = 2^k, k > 0$.

Proof. For $b > a$, let $\{0, 1, 2, \cdots, b\}$ and $\{0, 1, 2, \cdots, a\}$ be two subsets of $\{0, 1, 2, \cdots, n - 1\}$. This latter complete sequence cannot be a sequencing if

$$\sum_{i=0}^{b} i - \sum_{i=0}^{a} i = cn \equiv 0 \bmod n.$$

From Eq. (1), $c \leq s - 1$, that is, $\frac{n(n-2)}{2}$ is the largest partial sum which is an integral multiple of n.

$$\sum_{i=0}^{b} i - \sum_{i=0}^{a} i = \frac{b(b+1)}{2} - \frac{a(a+1)}{2} = \frac{(b-a)(b+a+1)}{2}$$

Let $b - a = d \leq n - 1$ and assume that for some $c \leq s - 1, (b-a)(b+a+1) = 2cn$, then:

$$b + a + 1 = \frac{2cn}{d} \text{ and } b = \frac{cn}{d} + \frac{d-1}{2} \tag{3}$$

If b is an integer less than n, $\{0, 1, 2, \cdots, n - 1\}$ is not a sequencing.

Case 1. $n = 2^k$

$d \leq n - 1$. If d is odd, and $d \mid c$, then $\frac{cn}{d} \geq n$. If $d \nmid c$, then $\frac{cn}{d}$ is not an integer but $\frac{d-1}{2}$ is an integer. Thus, b is not an integer. If d is even, then $\frac{d-1}{2}$ can be expressed as an integer $+\frac{1}{2}$; however, since $d < n$, either $\frac{cn}{d}$ is an integer or an integer plus a fraction with an odd denominator. Again, b is not an integer and so $\{0, 1, 2, \cdots, n - 1\}$ is a sequencing.

Case 2. $n = e2^l$ where e is odd, $3 \leq e \leq \frac{n}{2}, l \geq 1, n \geq 6$.

Let $d = e$ and $c = 1$. Then $b = \frac{cn}{d} + \frac{d-1}{2} = 2^l + \frac{e-1}{2}$ which is an integer.

We can write:

$$2^l = 2 + \alpha \text{ for } \alpha \geq 0. \quad e = 2 + \beta \text{ for } \beta \geq 1$$
$$(2 + \alpha) + (2 + \beta) = 4 + \alpha + \beta$$
$$(2 + \alpha)(2 + \beta) = 4 + 2\alpha + 2\beta + \alpha\beta > 4 + \alpha + \beta$$

Thus:

$$b = 2^l + \frac{e-1}{2} < 2^l + e < e2^l = n$$

so that $\{0, 1, 2, \cdots, n-1\}$ is not a sequencing.

Another aid in generating quick trickle permutations is to note that if $\{a_0, a_1, a_2, \cdots, a_{n-1}\}$ is a sequencing, then so is $\{a_0, a_{n-1}, a_{n-2}, \cdots, a_2, a_1\}$; and, conversely, if one is not, the the other is not. This is simply because addition modulo n is associative: $\sum_{i=b}^{c} a_i = \sum_{i=c}^{b} a_i$.

There are some patterns which occur in a group of permutations. Since $n = 2s$, $a_i = s$ for some index i in the sequencing corresponding to a quick trickle permutation g. For $k < n$ and GCD $(k, n) = 1$ the permutation and its kth power can be written:

$$g = (1, c_1, c_2, \cdots, c_i, \cdots, c_{n-1})$$
$$g^k = (1, d_1, d_2, \cdots, d_i, \cdots, d_{n-1})$$

$c_k = d_1, c_{2k} = d_2$ etc. In general, if $c_i = d_l$ then $kl \equiv i \bmod n$, and similarly if $c_j = d_m$.

The spacing between pairs of numbers $c_i \rightarrow c_j$, is $(j-i)$ in g and for $d_l \rightarrow d_m$ is $(m-l)$ in g^k. $(j-i) \equiv (m-l)k \bmod n$. In particular, the spacing a_i from $i \rightarrow i+1$ in g becomes a_i' in g^k where $a_i \equiv ka_i' \bmod n$. Starting with the basic permutation g, and the associated sequencing $\{0, a_1, a_2, \cdots, a_{n-1}\}$ the corresponding sequencing in g^k is $\{0, ta_1, ta_2, \cdots, ta_{n-1}\}$ where $kt \equiv 1 \bmod n$, that is, $t < n$ is the integer such that $(g^k)^t \equiv g \bmod n$. t is relatively prime to n. t is necessarily odd; so $t = 2d + 1$ where $d \geq 0$. Consider the number in the sequencing $\{a_i\}$, $a_j = s = \frac{n}{2}$. $a_j' = ta_j = (2d+1)\frac{n}{2} = dn + \frac{n}{2} \equiv s \bmod n$. This proves that following:

Proposition 3. In a group of quick trickle permutations, the sequencings (corresponding to maximal permutations) all have $a_j = \frac{n}{2}$ in the same position.

Definition. A sequencing corresponding to a quick trickle permutation for which $a_s = s$, is called symmetric if $a_i + a_{n-i} \equiv 0 \bmod n$. If $a_i + a_{s+i} \equiv 0 \bmod n$, it is called antisymmetric.

Note that $a_i + a_{n-i} \equiv 0 \bmod n$ is not a sufficient condition for a sequencing. As shown in Prop. 2, although $0, 1, 2, \cdots, \frac{n}{2}, \cdots, n-1$ has this property, it is not a sequencing if $n \neq 2^k$.

Proposition 4. (I) In the symmetric case, the sequencing corresponding to the inverse permutation is

$$\{0, a_{n-1}, \cdots, a_{s+1}, a_s, a_{s-1}, \cdots, a_2, a_1\}.$$

(II) In the antisymmetric case, the sequencing corresponding to the inverse permutation is:

$$\{0, a_{s+1}, \cdots, a_{n-1}, a_s, a_1, \cdots, a_{s-1}\}.$$

Proof. The sequencing corresponding to the inverse, as shown in the proof of Prop. 3, is $\{0, (n-1)a_1, (n-1)a_2, \cdots, (n-1)a_{n-1}\}$. In (I) $a_1 + a_{n-1} = na_1 \equiv 0$ mod n, so $(n-1)a_1 = a_{n-1}$ etc. In (II) $a_1 + a_{s+1} = na_1 \equiv 0$ mod n, so $(n-1)a_1 = a_{s+1}$ etc.

7 Graphic Representation of Quick Trickle Permutations

A quick trickle permutation can be represented as a directed graph. For a permutation of order $n = 2s$, the vertices of a regular n-side polygon, in some order, represent the permutation. Chords connecting numbers i to $i+1$, represent the spacings with values assigned as the number of edges subtended to the left, with respect to the direction of travel. By definition, for a quick trickle permutation, these chord lengths in terms of edges subtended must all be different. *Fig. 3* is an example for $n = 8$ for a permutation of the form of Eq. (2) in Section 6. In these graphs, there is a single path from each vertex to every successor. Clearly, if we can rearrange the chords, all of different length, so that we obtain another directed graph with unique paths, it will represent another quick trickle permutation.

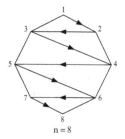

Fig. 3. $_8g_1 = (1\ 2\ 4\ 6\ 8\ 7\ 5\ 3)$

Using the convention mentioned above, the chord lengths or spacings marked by the arrow are, respectively, $\{0\ 1\ 6\ 3\ 4\ 5\ 2\ 7\}$. This is the sequencing corresponding to the permutation $_8g_1$. Clearly, any transformation to this graph which yields the same chord lengths will represent another quick trickle permutation and sequencing; for example, we can reverse the order of the chords or sequencing by following a symmetric, counterclockwise path like that shown in *Fig. 4*

In this case, we have obtained the inverse of the permutation in *Fig. 3*. The corresponding sequencing is: $\{0\ 7\ 2\ 5\ 4\ 3\ 6\ 1\}$, and the existence of the directed terrace is guaranteed by the fact that the above sequencing is the set of spacings in a permutation.

If we rotate the chord of length 4 in *Fig. 3* by $90°$, we obtain the new directed graph in *Fig. 5*.

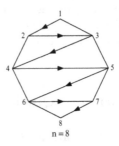

Fig. 4. $_8g_1^{-1} = {_8g_1^7} = (1\,3\,5\,7\,8\,6\,4\,2)$

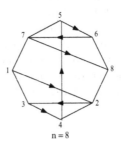

Fig. 5. $_8g_2 = (1\,7\,5\,6\,8\,2\,4\,3)$

In this case, the chord lengths, in order, or the sequencing is: $\{0\,5\,2\,7\,4\,1\,6\,3\}$. Note that this represents a rotation of blocks around $\frac{n}{2} = 4$ in the sequencing in *Fig. 3, i.e.*:

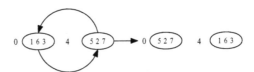

$_8g_2$ is not a power of $_8g_1$ and, thus, a second group of quick trickle permutations is determined. $_8g_2^5 = (1\,2\,5\,3\,8\,7\,4\,6)$ and $_8g_1 = (1\,2\,4\,6\,8\,7\,5\,3)$, generate two disjoint groups of quick trickle permutations.

The above example used a symmetric sequencing and generated two groups of quick trickle permutations. Next we consider a case in which the sequencing is asymmetric. The QTP is $_8h_i = (1\,2\,7\,4\,8\,6\,5\,3)$ and the corresponding sequencing is $\{1\,6\,4\,3\,7\,5\,2\}$.

This quick trickle permutation can be used to generate other groups of permutations. First of all, the diameter extending from vertex 3 to 4 in *Fig. 6* can be rotated $45°$ to connect the original vertices 1 and 8. The result is shown in *Fig. 7*.

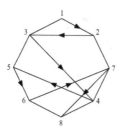

Fig. 6. Complete modification to **Fig. 3.** $_8h_1$ = (1 2 7 4 8 6 5 3), $\{a_i\}$ = $\{0\ 1\ 6\ 4\ 3\ 7\ 5\ 2\}$.

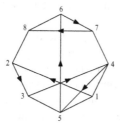

Fig. 7. Modification of **Fig. 6** by rotation of the diameter. $_8h_3$ = (1 5 3 2 8 6 7 4), $\{a_i\} = \{0\ 3\ 7\ 5\ 2\ 4\ 1\ 6\}$.

This is independent of the previous quick trickle permutations in the sense that it is not a power of any of them. As shown in *Fig. 6*, $_8h_1$ can also be modified by reversing the order of the corresponding sequencing. This is shown in *Fig.8*.

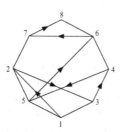

Fig. 8. Sequencing reversed from **Fig. 6.** $_8h_4$ = (1 5 2 7 8 6 4 3), $\{a_i\}$ = $\{0\ 2\ 5\ 7\ 3\ 4\ 6\ 1\}$.

Once again, this is not a power of a previous quick trickle permutation. Finally, we can rotate the diameter, leaving the other chords unchanged to obtain the directed graph in *Fig. 9*.

If one uses only modifications of a directed graph, that is, mirror image and rotation of the diameter or chord subtending four edges, then permutations

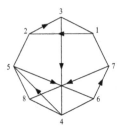

Fig. 9. Modification of **Fig. 8.** $_8h_2 = (1\,7\,6\,4\,8\,5\,2\,3), \{a_i\} = \{0\,6\,1\,4\,2\,5\,7\,3\}$.

$_8h_{1,8}\,h_{2,8}\,h_3$, and $_8h_4$ form an equivalence class of four groups, while $_8g_1$ and $_8g_2$ form an equivalence class of two groups.

There is a clear distinction between the directed graphs of *Fig. 3* and *4* on the one hand and those in *Fig. 6, 7* and *8* on the other hand. The former are symmetric about the diameter or chord of length $s = \frac{n}{2} = 4$; the latter are asymmetric around this diameter. The corresponding sequencings, omitting the initial zero, are similarly symmetric or asymmetric. Antisymmetric sequencings occur only for $n \geq 10$. As shown above, interchanging the blocks minus the zero, to the left and right of s in the sequencing produces another sequencing.

Simultaneously, reversing the order of these two blocks also produces another sequencing. This is shown schematically as follows:

$$0\{\text{block } L\}s\{\text{block } R\}$$
$$0\{\text{block } L^{-1}\}s\{\text{block } R^{-1}\}$$
$$0\{\text{block } R\}s\{\text{block } L\}$$
$$0\{\text{block } R^{-1}\}s\{\text{block } L^{-1}\}$$
where block $L = \{a_1, a_2, \ldots, a_k\}$, block $L^{-1} = \{a_k, \ldots, a_2, a_1\}$
block $R = \{a_{k+2}, \ldots, a_{n-1}\}$, block $R^{-1} = \{a_{n-1}, \ldots, a_{k+2}\}$
and s is in position $k + 1$.

In the symmetric case, $k + 1 = s$, that is s is in the middle of the nonzero numbers $\{a_i\}$ with block of $s - 1$ numbers in each of the blocks to the left and right of s. In that case, the second sequencing is in the reverse order of the first and the corresponding permutation is the inverse of the permutation corresponding to the first sequencing. There is a similar correspondence between the third and fourth sequencings.

A symmetric quick trickle permutation generates an equivalence class of two permutation groups. In the antisymmetric case, the left and right blocks respective to s in the sequencing are interchanged in the sequencing corresponding to the inverse permutation (see Prop. 4). In the asymmetric case, inverting the order of the numbers within the left and right blocks does not reverse the order of the sequencing and does not, correspondingly, generate the inverse of the first permutation. An asymmetric quick trickle permutation generates an equivalence class of four permutation groups.

8   Some Remarks and Observations

Searching for quick trickle permutations and sequencings is a first olive out of the bottle problem, that is, finding one yields two to four groups as shown above. Some cut and try processes are reasonably fruitful, such as, making minor modifications including expansion of existing permutations, sequencings and directed graphs. However, one deterministic method of generating symmetric QTP's is extrapolation, that is doubling a sequencing by using a size n sequencing as the left (or right) side of a new size $2n$ sequencing.

The number of groups of QTP's and sequencings for n_{10} is shown below. The ordinate is the block size $n = 2s$. The abscissa is the location of $a_i = s$ in the sequencing relative to the midpoint of the set of $n - 1$ nonzero numbers. This is the same as the location of the diameter in the directed graph.

	$s-4$	$s-3$	$s-2$	$s-1$	s	$s+1$	$s+2$	$s+3$	$s+4$
$n=4$				0	1	0			
$n=6$			0	0	2	0	0		
$n=8$		0	0	2	2	2	0	0	
$n=10$	0	12	8	8	16	8	8	12	0

Fig. 10.

References

1. Bailey, R.A.: Quasi-Complete Latin Squares; Construction and Randomization. Journal of the Royal Statistical Society B446, 323–334 (1984)
2. Buckley, F., Harvey, F.: Distance in Graphs. Addison-Wesley, Reading (1989)
3. Dénes, J., Keedwell, A.D.: Latin Squares - New Developments in the Theory and Applications. North-Holland, Amsterdam (1991)
4. Golomb, S.W., Taylor, H.: Tuscan Squares – a New Family of Combinatorial Designs. In: Ars Combinatoria, Waterloo, Canada, vol. 20-B, pp. 115–132 (1985)
5. Song, H.Y., Golomb, S.W.: Generalized Welch-Costas Sequences and Their Application to Vatican Arrays. In: Contemporary Mathematics, vol. 168, pp. 341–351. American Mathematical Society, Providence, RI (1994)
6. Denning, D.E.R.: Cryptography and Data Security. Addison-Wesley, Reading (1983)

Double Periodic Arrays with Optimal Correlation for Applications in Watermarking

Oscar Moreno[1] and José Ortiz-Ubarri[2]

[1] Computer Science Department
University Of Puerto Rico
Río Piedras Campus
moreno@uprr.pr
[2] High Performance Computing facility
University of Puerto Rico
jose.ortiz@hpcf.upr.edu

Abstract. Digital watermarking applications require constructions of double-periodic matrices with good correlations. More specifically we need as many matrix sequences as possible with both good auto- and cross-correlation. Furthermore it is necessary to have double-periodic sequences with as many dots as possible.

We have written this paper with the specific intention of providing a theoretical framework for constructions for digital watermarking applications.

In this paper we present a method that increases the number of sequences, and another that increases the number of ones keeping the correlation good and double-periodic. Finally we combine both methods producing families of double-periodic arrays with good correlation and many dots. The method of increasing the number of sequences is due to Moreno, Omrani and Maric. The method to increase the number of dots was started by Nguyen, Lázló and Massey, developed by Moreno, Zhang, Kumar and Zinoviev, and further developed by Tirkel and Hall. The very nice application to digital watermarking is due to Tirkel and Hall.

Finally we obtain two new constructions of Optical Orthogonal Codes: Construction A which produces codes with parameters $(n, \omega, \lambda) = (p(p-1), \frac{p^2-1}{2}, \lceil \frac{p(p+1)}{4} \rceil)$ and Construction B which produces families of code with parameters $(n, \omega, \lambda) = (p^2(p-1), \frac{p^2-1}{2}, \lceil \frac{p(p+1)}{4} \rceil)$ and family size $p+1$.

Keywords: double-periodic, correlation, watermark, sequences, matrix, array.

1 Background

Sequences with good auto- and cross-correlation have been studied by our group for their applications in frequency hopping radar and sonar, and communications, and more recently in digital watermarking.

S.W. Golomb et al. (Eds.): SSC 2007, LNCS 4893, pp. 82–94, 2007.

Costas and sonar sequences were respectively introduced in [3] and [5] to deal with the following fundamental problem:

"We have an object which is moving towards (or away from) us and we want to determine the distance and velocity of that object."

The solution to the problem makes use of the Doppler effect. Doppler observed that when a signal hits a moving object its frequency changes in direct proportion to the velocity of the moving object relative to the observer. In other words, if the observer sends out a signal towards a moving target, the change between the frequency of the outgoing and that of the returning signal will allow him to determine the velocity of the target, and the time it took to make the round trip will allow him to determine the distance.

In a frequency hopping radar or sonar system, the signal consists of one or more frequencies being chosen from a set $\{f_1, f_2, \ldots, f_m\}$ of available frequencies, for transmission at each of a set of $\{t_1, t_2, \ldots, t_n\}$ of consecutive time intervals. For modeling purposes, it is reasonable to consider the situation in which $m = n$, and where

$$\{f_1, f_2, \ldots, f_n\} = \{t_1, t_2, \ldots, t_n\} = \{1, 2, \ldots, n\}$$

(we will call this last $m = n$ case, a Costas type, and the general case sonar type).

Such a Costas signal is conveniently represented by a $n \times n$ permutation matrix A, where the n rows correspond to the n frequencies, the n columns correspond to the n time intervals, and the entry a_{ij} equals 1 if and only if frequency i is transmitted in time interval j. (Otherwise, $a_{ij} = 0$)

When this signal is reflected from the target and comes back to the observer, it is shifted in both time and frequency, and from the amounts of these shifts, both range and velocity are determined. The observer finds the amounts of these shifts by comparing all shifts (in both time and frequency) of a replica of the transmitted signal with the actual received signal, and finding for which combination of time shift and frequency shift the coincidence is greater. This may be thought of as counting the number of coincidences between 1's in the matrix $A = (a_{ij})$ with 1's in a shifted version A^* of A, in which all entries have been shifted r units to the right (r is negative if there is a shift to the left), and s units upward (s is negative if the shift is downward). The number of such coincidences, $C(r, s)$, is the two-dimensional auto-correlation function between A and A^*, and satisfies the following conditions:

$$C(0,0) = n$$

$$0 \leq C(r, s) \leq n \text{ except for } r = s = 0$$

(This conforms to the assumption that the signal is 0 outside the intervals $1 \leq f \leq n$ and $1 \leq t \leq n$)

If we have another Costas type of signal represented by a matrix $B = (b_{ij})$, we can similarly define the two-dimensional cross-correlation function by substituting A^* by B^* in the above definition.

In the real world, the returning signal is always noisy. The two-dimensional auto-correlation function, $C(r, s)$, is also called the ambiguity function in radar and sonar literature, and should be thought of as the total "coincidence" between the actual returning noisy signal and the shift of the ideal transmitted signal by r units in time and s units in frequency. Among the 2^{n^2} matrices of 0's and 1's of order n, there are $n!$ permutation matrices, and some of these are not very good as signal patterns for radar and sonar. For example, the $n \times n$ identity matrix I_n can be shifted one unit up and one unit left, and will then produce $n - 1$ coincidences with the original matrix. For large values of n and a noisy environment, the signal pattern I_n would most certainly produce spurious targets, shifted an equal number of units in both time and frequency from the real target.

At a minimum, there is a shift of $A = (a_{ij})$ which will make any of the n 1's land on any of the $n - 1$ remaining 1's, so we know that

$$\min C(r, s) = 1$$

$$\max C(r, s) \geq 1$$

$$\text{for all "codes"} (r, s) \neq (0, 0)$$

where $C(r, s)$ is the ideal ambiguity function of the permutation matrix itself. Consequently J.P. Costas [3] defined the ideal $n \times n$ permutation matrices (which we will call here Costas sequences) as those for which

$$\max C(r, s) = 1 \text{ when } (r, s) \neq (0, 0)$$

By hand computation, he found examples of such matrices for all $n \leq 12$, but was unable to find an example for $n = 13$, and was tempted to conclude that these patterns "die out" beyond $n = 12$.

In the general sonar case, n signals are sent out with frequencies ranging from 1 to m, at times ranging from 1 to n. Once the whole pattern of signals has returned, the velocity and the distance of the object can be determined as mentioned before. For sonars you must have exactly a 1 in every column but the rows can have multiple 1's or they can be empty of 1's. The problem in sonars (see [7]) is for any n obtain the largest possible m.

It has been proven in [4] that for $n > 3$ there are no two different Costas sequences with the same ideal property in their cross-correlation as that they have in their auto-correlation. Since for the case of multiple targets we need sets of sequences with good auto- and cross-correlation properties we had therefore settled for constructing sets of sequences with nearly ideal properties, or in other words cross-correlation 2.

In spread spectrum communications the data sent in a communication channel is spread to avoid its interception and channel jamming; and in modern communications like CDMA for multiple access in wireless and optical communication. Using codes with good auto- and cross-correlation a message sent in a communication channel can then be easily recovered in the other side of communication and furthermore using codes with good cross-correlation allows communication of multiple users limiting signal interference.

Recently sequences with good auto- and cross-correlation have been used in the area of digital watermarking because they make watermarks more difficult to detect, damage or remove from a digital medium. The idea is similar to the spread spectrum communications where a secret message is spread into a channel in order to make it more difficult to be intercepted or removed.

A watermark is an array or a sum of arrays that can carry information. This array is added to a medium in order to make it difficult to perceive. The watermark is recovered by calculating the watermark correlation with the watermarked medium. Families of arrays with perfect or near to perfect auto- and cross-correlation allow the addition of multiple arrays to increase information capacity (or multiple users) and watermark security.

In previous work, families of Costas and sonar arrays have been studied in the area of digital watermarking. The Moreno-Maric construction [12], which generates families of double periodic arrays, was used by Tirkel and Hall in particular because of their near to perfect correlation (2) and the size of its families. In order for watermarks to be more effective it must have many dots. Also it is necessary to have as many sequences with low cross-correlation as possible to combine them and increase the watermark information capacity. The method of using periodic-sequences to replace columns of matrices in order to increase the number of dots in double-periodic sequences was introduced in [10], was previously used by our group [8] and also used by Tirkel and Hall [11] in the area of watermarking. Recently Moreno, Omrani and Maric [9] presented a new construction of double periodic sequences with perfect auto- and cross-correlation. In this work we will use this new construction to generate families of matrices that can be used for digital watermarking.

2 Method to Increase the Number of Sequences without Increasing the Original Correlation Value

Moreno, Omrani and Maric showed how to construct new families of sonars and extended Costas arrays, from a Welch Costas array $(p \times (p-1))$, with auto- and cross-correlation 1. The Welch Costas arrays are constructed as follows:

Welch Construction. Let α be a primitive root of an odd prime p.
Then the array with

$$\alpha_{k,j} = 1, 1 \leq k, j \leq p-1$$

if and only if

$$j \equiv \alpha^k (\text{mod } p), 1 \leq k \leq p-1,$$

otherwise $\alpha_{k,j} = 0$, is a Costas array.

This construction is the first construction of multiple target sonars with perfect auto- and cross-correlation properties. Multiple target arrays are families of arrays used in radar and sonar that are sent to different targets. When the echoes are received the low cross-correlation of the arrays is used to distinguish the distance and velocity of each target. In the case of watermarking we call these arrays multiple user arrays, because instead of using the arrays to distinguish targets we use them to distinguish users.

2.1 OOC, DDS, and Double Periodic Arrays of Families

An (n, ω, λ) Optical Orthogonal Code (OOC) C where $1 \leq \lambda \leq \omega \leq n$, is a family of $\{0,1\}$-sequences of length n and Hamming weight ω satisfying:

$$\sum_{k=0}^{n-1} x(k)y(k \oplus_n \tau) \leq \lambda \qquad (1)$$

whenever either $x \neq y$ or $\tau \neq 0$. We will refer to λ as the maximum correlation parameter, and Φ as the family size.

A (k, v)-Distinct Difference Set (DDS) [1] is a set $\{c_i | 0 \leq i \leq k - 1\}$ of distinct integers such that the $k(k-1)$ differences $c_i - c_j$ where $i \neq j$ are distinct modulo v.

By a (v, k, t)-DDS, we mean a family $(B_i | i \in I, t = |I|)$ of subsets of $\mathbb{Z}_v$ each of cardinality k, such that among the $tk(k-1)$ differences $(a - b | a, b \in B_i; a \neq b; i \in I)$ each nonzero element $g \in \mathbb{Z}_v$ occurs at most once. This notion of a (v, k, t)-DDS is a more recent generalization of the earlier concept of a (k, v)-DDS. A (k, v)-DDS is a (v, k, t)-DDS with parameter $t = 1$.

Lemma 1. *There is a one to one onto correspondence between the set of (n, ω, λ)-OOCs and the set of (v, k, t)-DDSs when $\lambda = 1$ with $n = v$, $k = \omega$ and $\Phi(n, \omega, 1) = t$, and $\Phi(n, \omega, 1)$ is the family size of the OOCs.*

Proof. The incidence vectors associated to the subsets comprising a (v, k, t)-DDS can be seen to form an (n, ω, λ)-OOC of size t with parameters $n = v, w = k$, and $\lambda = 1$. Conversely, given an OOC and a maximal set of cyclically distinct representatives drawn from the code, one obtains a DDS by considering the support of these vectors. Thus, the concept of (v, k, t)-DDS is precisely the same as that of an OOC with $\lambda = 1$.

Let $A = [A(i, j)]$ and $B = [B(i, j)]$ be $r \times s$ matrices having 0,1 entries where r and s are relatively prime. We now have the following definition:

Definition 1. *The double-periodic cross-correlation between A and B is an integer valued function for a change of value (a, b) a in the row and b in the column. In other words the function varies for all (a, b) a less than the first value and b less than the second value of the double periodicity. The function $C(a, b)$ is a integer function defined as follows:*

$$\sum_{i=0}^{r-1} \sum_{j=0}^{s-1} A(i \oplus_r \alpha, j \oplus_s \tau)B(i, j) \leq C(a, b) \qquad (2)$$

for any $\alpha \leq r, \tau \leq s$, where $\oplus_m$ denotes addition modulo m the smallest such $C(a,b)$ is the correlation. Auto-correlation is the same with $A = B$. Let $a(.)$ and $b(.)$ be the sequences of length rs associated with the matrices A and B respectively via the Chinese Remainder Theorem, $a(L) = A(L(\mod r), L(\mod s))$ and similarly $b(L) = B(L(\mod r), L(\mod s))$ for all L, $0 \leq L \leq rs - 1$.

Definition 2. *Bound on the correlation.*

$$max\ C(a,b) \leq \lambda \text{ when } (a,b) \neq (0,0) \tag{3}$$

From the previous definitions we obtain the following theorem:

Theorem 1. *The collection of one-dimensional periodic auto- and cross-correlation values of a family of sequences of length rs is precisely the same as the set of two dimensional double-periodic auto- and cross-correlation values of $r \times s$ matrices associated with these sequences via the residue map, whenever r and s are relatively prime.*

Corollary 1. *The concept of an OOC with auto- and cross-correlation λ is the same as that of a double-periodic multi-target arrays with auto- and cross-correlation λ.*

1) MZKZ Construction A: When m is a divisor of $p-1$, $m|(p-1)$, and p is a prime, the construction of an $(n = mp, w = m, \lambda = 1)$, $\Phi = \frac{p-1}{m}$ OOC (Construction A in Moreno et al [8]) yields a $(v = mp, k = m, \frac{p-1}{m})$-DDS for any $m|(p-1)$. This construction is optimal with respect to the Johnson Bound [6] on the cardinality of a constant weight binary code when $p > 3$ and $m = p - 1$. The construction is given for $m = p - 1$ in the following:

If we choose any degree one polynomial $f(x)$ over $\mathbb{F}_p$, and fill out the elements of a $p \times (p - 1)$ matrix M with the following rule:

$$M(i,j) = \begin{cases} 1, & \text{if } f(\alpha^j) = p - 1 - i \\ 0, & \text{otherwise} \end{cases} \tag{4}$$

where α is a primitive element of F_p, then the resulting M matrix has one 1 per column and has the double-periodic auto-correlation property. If we apply the Chinese Remainder Theorem to the matrix M we will end up with an OOC sequence μ of length $p(p - 1)$:

$$\mu(l) = M(l \mod (p), l \mod (p - 1)) \tag{5}$$

2) A New Family of OOC's: M.J. Colbourn and C.J. Colbourn [2] proposed two recursive constructions for cyclic BIBD's. Their Construction A was generalized [13] to form DDS recursively. The following is an easy generalization of Colbourn construction B:

Construction B: Given a (vk, k, t)-DDS, $(vk = 0(\mod k))$ if $gcd(r, (k-1)!) = 1$, then a (vkr, k, rt)-DDS may be constructed as follows. For each $D = \{0, d_1, \ldots, d_{k-1}\}$, take the r difference sets $\{0, d_1 + ikv, d_2 + 2ikv, \ldots, d_{k-1} + (k - 1)ikv\}$,

$0 \leq i < r$, with addition performed modulo vkr. If furthermore, there exists an (rk, k, t')-DDS D', then a $(vkr, k, rt + t')$-DDS can be constructed by adding the t' difference sets $\{0, vs_1, \ldots, vs_{k-1}\}$ for each $D'_i = \{0, s_1, \ldots, s_{k-1}\}$ of $D' = \{D'_i | 1 \leq i \leq t'\}$.

A proof of the above is not included (the proof is similar to the one in [2]). In Lemma 2 we will prove the special case that interests us in this paper.

Construction CMZKZ: Applying construction B recursively to MZKZ family A construction, we obtain a $(p^i(p-1), p-1, 1)$-OOC of size $p^{i-1}+p^{i-2}+\cdots+p+1$. This OOC is not optimal with respect to the Johnson Bound [6].

Lemma 2. *In the $(p^i(p-1), p-1, 1)$-OOC of the above construction, all residues occur exactly once except multiples of $p-1$ and p^i.*

Proof. In the base OOC all the residues occur except multiples of p and $p-1$. Now applying the recursive construction to the $(p(p-1), p-1, 1)$ base OOC, in the resulting $(p^2(p-1), p-1, 1)$-OOC all the multiples of residues present in the base OOC will be present in addition to the multiples of p times the residues of the base OOC. So in the new OOC the multiples of $p-1$ do not occur. In addition since the multiples of p were not present in the base residues so in the new OOC the multiples of p^2 also do not occur.

We can use the same proof inductively to prove that in $(p^i(p-1), p-1, 1)$ all the residues occur exactly once except the multiples of p^i and $p-1$.

2.2 Two New Multiple Target Families for Extended Costas and for Sonar Arrays

Using the Chinese Reminder Theorem and Theorem 1 of Section 2.1, since p^i is relatively prime to $p-1$ we obtain:

Construction 1(V): From Section 2.1 we obtain a family of $p^2 \times (p-1)$ sonar arrays with family size of $p+1$ with auto- and cross-correlation 1.

Construction 2(V): A family of $p^i \times (p-1)$ sonar arrays with family size of $p^{i-1} + p^{i-2} + \cdots + 1$ with auto- and cross-correlation 1.

Construction 1(H): From section 2.1 we obtain a family of $(p-1) \times p^2$ extended Costas arrays with family size of $p+1$ with auto- and cross-correlation 1.

Construction 2(H): A family of $(p-1) \times (p^i)$ extended Costas arrays with family size of $p^{i-1} + p^{i-2} + \cdots + 1$ with auto- and cross-correlation 1.

Example 1. An example to generate the family of Construction 1(V). Start with a Welch array of Figure 1 2,4,3,1. Now notice that $(0,2)$ corresponds to 12 using the Chinese Remainder Theorem since $12 \equiv 0 \mod (4)$ and $12 = 2 \mod (5)$. Also $(1, 4) \rightarrow 9$, $(2, 3) \rightarrow 18$, and $(3, 1) \rightarrow 11$. Where in $(x, y) \rightarrow z$, x is the value of the column, y is the value of the row, and z is the Chinese Remainder for (x, y). Applying the Chinese Remainder Theorem to the Welch array we obtain the OOC D:

$$D = \{9, 11, 12, 18\}$$

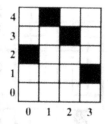

Fig. 1. 5x4 Welch Costas

which is equivalent to D':

$$D' = \{9, 11, 12, 18\}$$

From D' using our construction B we obtain the 6 arrays D_1, D_2, D_3, D_4, D_5 and D_6 as follows: (See section 2.1):

$$D_1 = \{0, 2, 3, 9\} \text{ for } i = 0$$

$$D_2 = \{0, 22, 43, 69\} \text{ for } i = 1$$

$$D_3 = \{0, 29, 42, 83\} \text{ for } i = 2$$

$$D_4 = \{0, 23, 62, 89\} \text{ for } i = 3$$

$$D_5 = \{0, 49, 63, 82\} \text{ for } i = 4$$

Now we multiply D' by 5:

$$D_6 = \{0, 10, 15, 45\}$$

Finally apply the Chinese Reminder Theorem again to each D_i to construct the family of 25×4 sonars arrays of size 6. I.E. To construct sonar S_i for each element $d \in D_i$, calculate $s = (d \mod 4, d \mod 25)) \in S_i$. See Figure 2.

3   Method to Increase the Number of Dots

Previous work [10] [8] [11] describes how to increase the number of dots in a double-periodic matrix sequence using periodic shift sequences. Tirkel and Hall applied this method with the Moreno-Maric construction [12] to create new matrices with good auto- and cross-correlation.

Method A: consists in replacing the columns of a double-periodic matrix W with a cyclically shifted periodic sequence s with the size of the columns of the matrix (See Figure 3(b)). For each column j in W, find the row i where $W_{i,j} = 1$, construct s' such that s' is equal to s cyclically shifted i units, and replace the column j with s'. Figure 3 is an example of a double-periodic Welch 7×6

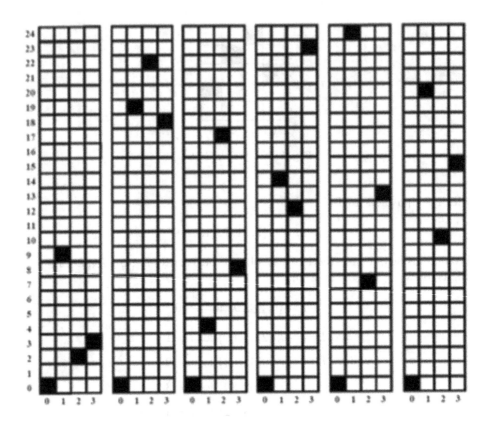

Fig. 2. 25x4 Moreno-Omrani-Maric sonars family

matrix sequence $(0,3), (1,2), (2,6), (3,4), (4,5), (6,1)$ with the columns replaced by a binary m-sequence $(0,0,1,1,1,0,1)$ of size 7.

Proof of the following theorem can be done following the techniques used by Nguyen, Lázló and Massey in [10].

Theorem 2. *Method A applied to a Welch array of size $p(p-1)$ using a Legendre sequence as a column produces OOCs with parameters $(n, \omega, \lambda) = (p(p-1), \frac{p^2-1}{2}, \lceil \frac{p(p+1)}{4} \rceil)$. These codes are asymptotically optimum.*

4   New Matrix Construction for Watermarking

In section 2 we showed how to construct families of double-periodic sequences with perfect correlation. This property is very useful in digital watermarking because it reduces the number of false positives in watermark detection. In section 3 we explained how to increase the number of dots in a double-periodic sequence. In the next subsection we will use the Moreno-Omrani-Maric family construction to construct new families of matrices which are more efficient for watermarking by increasing the number of dots in the Moreno-Omrani-Maric construction.

4.1 Method to Increase the Number of Dots and the Number of Sequences with Optimal Correlation

We construct a family of matrices from a Welch Costas array using column sequences and applying the Moreno-Omrani-Maric construction.

Method B: First we generate a Welch Costas array, then we replace the columns with a suitable cyclically shifted periodic sequence to increase the number of dots (filled pixels in images) in the matrix, and finally we apply the Moreno-Omrani-Maric construction to generate the new family of size $p + 1$. (See example 2).

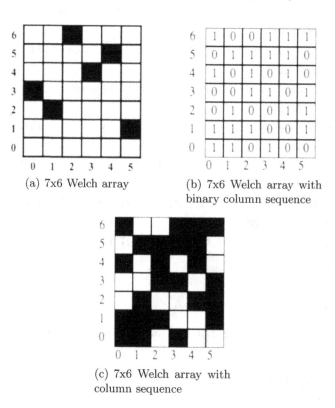

(a) 7x6 Welch array

(b) 7x6 Welch array with binary column sequence

(c) 7x6 Welch array with column sequence

Fig. 3. 7x6 Welch array with and w/o column sequence

Example 2. Start with the Welch array of Figure 3(a) with points $(0, 3), (1, 2),$ $(2, 6), (3, 4), (4, 5)$ and $(5, 1)$. Replace the columns of the matrix in that figure with the periodic sequence $0, 0, 1, 1, 1, 0, 1$ which is a binary m-sequence with auto-correlation 2 (See Figure 3(b)) to form the matrix in figure 3(c).

Now apply the Moreno-Omrani-Maric construction to the new matrix of size $p \times (p-1)$. The Chinese Reminder for the points in the new matrix are $(0,0) \rightarrow 0, (0,1) \rightarrow 36, (0,4) \rightarrow 18, (0,6) \rightarrow 6, (1,0) \rightarrow 7, (1,1) \rightarrow 1, (1,2) \rightarrow 37, (1,5) \rightarrow 19, (2,1) \rightarrow 8, (2,3) \rightarrow 38, (2,4) \rightarrow 32, (2,5) \rightarrow 26, (3,0) \rightarrow 21, (3,3) \rightarrow 3, (3,5) \rightarrow 33, (3,6) \rightarrow 27, (4,2) \rightarrow 16, (4,4) \rightarrow 4, (4,5) \rightarrow 40, (4,6) \rightarrow 34, (5,1) \rightarrow 29, (5,2) \rightarrow 23, (5,3) \rightarrow 17, (5,6) \rightarrow 41$.

From the Chinese Reminder Theorem we obtain D':

$$D' = \{0, 1, 3, 4, 6, 7, 8, 16, 17, 18, 19, 21, 23, 26, 27, 29, 32, 33, 34,$$
$$36, 37, 38, 40, 41\}$$

Now following the Moreno-Omrani-Maric construction we obtain from D':

$$D_1 = \{0, 1, 3, 4, 6, 7, 8, 16, 17, 18, 19, 21, 23, 26, 27, 29, 32, 33, 34,$$
$$36, 37, 38, 40, 41\}$$
$$D_2 = \{0, 16, 27, 38, 43, 59, 71, 82, 87, 102, 116, 125, 130, 145, 159, 174, 189, 202, 217,$$
$$233, 246, 260, 278, 289\}$$
$$D_3 = \{0, 16, 27, 38, 48, 63, 76, 85, 101, 113, 124, 133, 149, 162, 171, 186, 200, 209, 218,$$
$$236, 247, 256, 271, 285\}$$
$$D_4 = \{0, 16, 27, 38, 49, 65, 78, 88, 103, 117, 127, 143, 155, 166, 176, 194, 205, 216, 231,$$
$$244, 255, 270, 284, 293\}$$
$$D_5 = \{0, 16, 27, 38, 45, 60, 74, 83, 90, 105, 118, 134, 152, 163, 169, 185, 197, 208, 214,$$
$$229, 243, 259, 275, 288\}$$
$$D_6 = \{0, 16, 27, 38, 46, 61, 75, 92, 110, 121, 129, 144, 158, 167, 175, 191, 204, 211, 227,$$
$$239, 250, 258, 273, 286\}$$
$$D_7 = \{0, 16, 27, 38, 50, 68, 79, 91, 107, 120, 132, 147, 160, 172, 187, 201, 213, 228, 242,$$
$$251, 253, 269, 281, 292\}$$

And multiplying D' by 7:

$$D_8 = \{0, 7, 21, 28, 42, 49, 56, 112, 119, 126, 133, 147, 161, 182, 189, 203, 224, 231, 238,$$
$$252, 259, 266, 280, 287\}$$

Finaly apply the Chinese Reminder Theorem again to convert them to 6×49 matrices. (see Figure 4)

Theorem 3. *Method B applied to a Welch array of size $p(p-1)$ using a Legendre sequence as a column produces OOCs with parameters $(n, \omega, \lambda) = (p^2(p-1), \frac{p^2-1}{2}, \lceil \frac{p(p+1)}{4} \rceil)$ and family size $p+1$.*

In our example 2 we construct code sequences with $(n, \omega, \lambda) = (6 \times 49, 24, 14)$. These families of matrices can be used in digital watermarking because of their good cross-correlation. In our example the cross-correlation is 14 because of

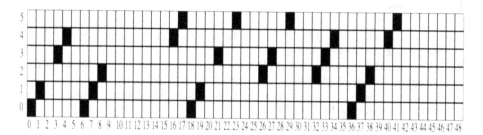

Fig. 4. 6x49 New matrix construction from Welch array 7x6 (rotated 90 degrees to the right)

the choice of the column sequence which auto-correlation is 2. The addition of arrays in the same family allows the watermark to carry more information. In the case of images the use of other methods like interlacing [12] can be also applied to extend families of matrices but increasing the correlation values. Also the selection of the column sequence and its correlation affects the new matrices' cross-correlation.

5   Conclusion

The Moreno-Omrani-Maric construction generates families with perfect auto- and cross-correlation. The matrices generated by this construction have few dots, and we showed a method to increase the number of dots in the matrices making them more effective for watermarking. We take advantage of the perfect correlation properties of the Moreno-Omrani-Maric construction to keep a low cross-correlation between matrices in the same family. In summary we showed a method to increase the number of matrix sequences and the number of dots in those matrices resulting in matrix sequences with good auto- and cross-correlation that can be used in digital watermarking.

We obtain two new constructions of Optical Orthogonal Codes: Construction A which produces codes with parameters $(n, \omega, \lambda) = (p(p-1), \frac{p^2-1}{2}, \lceil \frac{p(p+1)}{4} \rceil)$ and construction B which produces families of code with parameters $(n, \omega, \lambda) = (p^2(p-1), \frac{p^2-1}{2}, \lceil \frac{p(p+1)}{4} \rceil)$ and family size $p+1$.

Acknowledgments

We would like to thank Dr. Tirkel for introducing our group to the watermarking area. This work is partially supported by the High Performance Computing facility of the University of Puerto Rico under PR AABRE grant #P20RR16470 from the National Institute of Health and the Gauss Research Laboratory under SCORE grant #S06GM08102.

References

1. Coldbourn, C.J, Dinitz, J.H.: The CRC Handbook of Combinatorial Designs. CRC Press, Boca Raton, USA (1996)
2. Coldbourn, M.J., Coldbourn, C.J.: Recursive constructions for cyclic block designs. Journal of Statistical Planning and Inference 10, 97–103 (1984)
3. Costas, J.P.: Medium constraints on sonar design and performance. FASCON Convention Record, 68A–68L (1975)
4. Freedman, A., Levanon, N.: Any two n x n costas signal must have at least one common ambiguity sidelobe if n ¿ 3- a proof. In: Proceedings of the IEEE, vol. 73, pp. 1530–1531. IEEE Computer Society Press, Los Alamitos (October 1985)
5. Golomb, S.W., Taylor, H.: Two-dimensional syncronization patterns for minimum ambiguity. IEEE Trans. Information Theory IT-28, 600–604 (1982)
6. Johnson, S.M.: A new upper bound for error-correcting codes. IEEE Trans. on Information Theory IT(8), 203–207 (1962)
7. Moreno, O., Games, R.A., Taylor, H.: Sonar sequences from costas arrays and the best known sonar sequences with up to 100 symbols. IEEE Trans. Information Theory 39, 1985–1987 (1993)
8. Moreno, O., Zhang, Z., Kumar, P.V., Zinoviev, V.: New constructions of optimal cyclically permutable constant weight codes. IEEE Trans. Information Theory 41, 548–555 (1995)
9. Moreno, O., Omrani, R., Maric, S.V.: A new construction of multiple target sonar and extended costas arrays with perfect correlation. In: Proceedings of the 40th Annual Conference on Information Sciences and Systems, Princeton, NJ, USA (March 2006)
10. Nguyen, Q.A., Gyorfi, L., Massey, J.L.: Construction of binary constant-weight cyclic codes and cyclically permutable codes. IEEE Trans. Information Theory 38(3), 940–949 (1992)
11. Tirkel, A.Z., Hall, T.E.: Matrix construction using cyclic shifts of a column. In: Proceedings International Symposium on Information Theory, pp. 2050–2054 (September 2005)
12. Tirkel, A., Hall, T.: New matrices with good auto and cross-correlation. IEICE Trans. Fundam. Electron. Commun. Comput. Sci. E89-A(9), 2315–2321 (2006)
13. Zhi, C., Pingzhi, F., Fan, J.: Disjoint difference sets, difference triangle sets and related codes. IEEE Trans. Information Theory 38, 518–522 (1992)

Sequences for Phase-Encoded Optical CDMA

Reza Omrani[1], Pankaj Bhambhani[2], and P. Vijay Kumar[*,2]

[1] EE-Systems, University of Southern California,
Los Angeles, CA 90089-2565
`rezaomrani@gmail.com`
[2] ECE Department, Indian Institute of Science
Bangalore 560012
{`pankaj83,vijay`}`@ece.iisc.ernet.in`

Abstract. In phase-encoded optical CDMA (OCDMA) spreading is achieved by encoding the phase of signal spectrum. Here, a mathematical model for the output signal of a phase-encoded OCDMA system is first derived. This is shown to lead to a performance metric for the design of spreading sequences for asynchronous transmission.

Generalized bent functions are used to construct a family of efficient phase-encoding sequences. It is shown how M-ary modulation of these spreading sequences is possible. The problem of designing efficient phase-encoded sequences is then related to the problem of minimizing PMEPR (peak-to-mean envelope power ratio) in an OFDM communication system.

Keywords: Generalized bent function, optical code-division multiple access (OCDMA), Optical CDMA, phase-encoded optical CDMA, phase sequence, peak to average power ratio (PAPR), peak-to-mean envelope power ratio (PMEPR).

1 Introduction

There has been a recent upsurge of interest in applying Code Division Multiple Access (CDMA) techniques to optical networks [1,2,3].

There are two main approaches to data modulation and spreading in optical CDMA (OCDMA). The first approach, known as direct-sequence encoding [1], makes use of on-off-keying (OOK) data modulation and unipolar spreading sequences with good correlation properties. The spreading sequences used in these systems are called optical orthogonal codes (OOC) and these have been studied since 1990s [4]. There are many constructions and bounds on size of these codes in the literature. Algebraic constructions for families of OOCS can be found for instance, in [4,5,6,7,8,9,10,11,12].

The second OCDMA approach uses spectral encoding. In this method spreading is achieved by encoding of amplitude or phase of data spectrum[2,3]. The

---

[*] P. Vijay Kumar is currently on leave of absence from USC at the ECE Department, Indian Institute of Science, Bangalore. This work was supported in part, by the DARPA OCDMA Program under Grant No. N66001-02-1-8939.

spectral encoding OCDMA system is harder to implement in comparison with direct-sequence OCDMA. That is perhaps the reason why most studies on spectral encoding systems have an implementation focus. There are several variations of this method. The original papers in this field [2,3] suggested phase encoding of a coherent laser source. Because of the difficulty involved in generating coherent lasers, spectral amplitude encoding of non-coherent sources was subsequently suggested [13]. The current increasing demand for higher speeds has led to renewed interest in the use of coherent sources and spectral encoding [14,15,16].

As mentioned above, the main focus in spectrally-encoded OCDMA has been on their implementation; there is not much literature on the subject of spreading sequence design with the exception of a few results on spectral amplitude encoding. Most experimental results reporting on spectral phase-encoded OCDMA have assumed synchronous systems and the use of Walsh-Hadamard sequences as spreading sequences. As will be shown below, Walsh-Hadamard Sequences are indeed ideal for the synchronous case but quite unsuitable for asynchronous systems. Other papers in the literature have proposed the use of m-sequences or Gold sequences as spreading sequences but do not provide adequate justification for their use.

In this paper we first present a model of an asynchronous phase-encoded OCDMA system, and then identify a metric reflective of the amount of the cross-correlation (other-user interference) in the system. Based on this model, we formulate the sequence design problem. As will be shown, this problem is closely related to the PAPR and PMEPR problems in OFDM [17,18,19,20]. Finally, generalized bent functions [21] are used to construct efficient spreading sequences for an asynchronous phase-encoded OCDMA system. Furthermore, the same sequences are used to propose an M-ary modulation scheme for phase-encoded OCDMA.

2   System Model

The system that we are modelling in this paper is a phase encoding OCDMA system with coherent laser source. A diagram of this system with one transmitter and receiver is shown in Figure 1. The typical laser sources used for coherent transmission are mode locked lasers (MLL). The electrical field of a mode locked laser can be written as:

$$E_{MLL}(t) = e^{i\omega_0 t} \sum_{k=0}^{K-1} e^{ik(\Delta\omega)t} \tag{1}$$

In this equation K is the number of modes in the mode locked laser, and $\Delta\omega$ is the channel spacing between two consecutive modes in the mode locked laser.

The output of MLL is then passed through a phase encoder. In our model the phase encoder applies different phase shifts to different modes of MLL to spread it. Conventionally the phase masks used in this approach consists of only $\{0, \pi\}$ phase shifts. Recently, Stapleton et al. [22,23] show that using microdisk

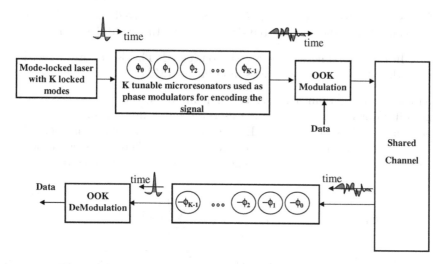

Fig. 1. Phase Encoding OCDMA system with Coherent Source

resonator technology any phase can be applied to the different modes of MLL. Considering this result no restriction on the choice of phases is considered in this paper. The output of the phase encoder will be of the form:

$$E_{Enc}(t) = e^{iw_0 t} \sum_{k=0}^{K-1} e^{i(k(\Delta w)t + \phi_k)} \tag{2}$$

where ϕ_k is the phase shift that encoder applies to kth mode of MLL. Upon OOK modulation with data bit d:

$$E_{Tr}(t) = d e^{iw_0 t} \sum_{k=0}^{K-1} e^{i(k(\Delta w)t + \phi_k)}, \quad d \in \{0,1\} \tag{3}$$

At the receiver, the phase decoder applies the inverse phase shift $-\phi_k$ to each mode k of the received signal:

$$E_{Dec}(t) = d e^{iw_0 t} \sum_{k=0}^{K-1} e^{i(k(\Delta w)t + \phi_k - \phi_k)}$$

$$= d e^{iw_0 t} \sum_{k=0}^{K-1} e^{ik(\Delta w)t} \tag{4}$$

which is the original signal in (1) modulated by data bit d. After the phase decoder a photo detector is used to detect the intensity of the received signal:

$$|E_{Dec}(t)|^2 = \left| d e^{iw_0 t} \sum_{k=0}^{K-1} e^{ik(\Delta w)t} \right|^2 = d \left| \sum_{k=0}^{K-1} e^{ik(\Delta w)t} \right|^2 \tag{5}$$

If we sample this signal at time $t = 0$ then the received signal will be dK^2 and we can retrieve the transmitted data d using a threshold detector.

Note 1. In this model no noise source is considered. This is because we wish to focus on the effect of multiple access interference (MAI).

When there is more than one user transmitting data, the receiver receives the superposition of the signals. Assume users m and n are transmitting data simultaneously and asynchronously. Each user uses its own phase encoder $\Phi^{(\ell)} = \{\phi_0^\ell, \phi_1^\ell, \cdots, \phi_{K-1}^\ell\}$ where $\ell \in \{m, n\}$. Let the time difference between user m and n be denoted by τ ($\tau = 0$ in a synchronous system). The received signal is given by

$$E_{Tr}^{(m)}(t) + E_{Tr}^{(n)}(t+\tau) = d^{(m)} e^{i\omega_0 t} \sum_{k=0}^{K-1} e^{i(k(\Delta\omega)t + \phi_k^{(m)})}$$

$$+ d^{(n)} e^{i\omega_0(t+\tau)} \sum_{k=0}^{K-1} e^{i(k(\Delta\omega)(t+\tau) + \phi_k^{(n)})} \tag{6}$$

The signal at the output of the phase decoder tuned to user m takes on the form :

$$d^{(m)} e^{i\omega_0 t} \sum_{k=0}^{K-1} e^{i(k(\Delta\omega)t)} + d^{(n)} e^{i\omega_0(t+\tau)} \sum_{k=0}^{K-1} e^{i(k(\Delta\omega)(t+\tau) + (\phi_k^{(n)} - \phi_k^{(m)}))} \tag{7}$$

The output of the photo detector of this receiver will be the square of the magnitude square of the above. As can be seen, there is multiple access interference(MAI) at the receiver output. Each transmitter-receiver pair is assumed to operate synchronously, and consequently, the receiver samples its output at time $t = 0$ to get:

$$A = \left| d^{(m)} K + d^{(n)} e^{i\omega_0 \tau} \sum_{k=0}^{K-1} e^{i(k(\Delta\omega)\tau + (\phi_k^{(n)} - \phi_k^{(m)}))} \right|^2 =$$

$$d^{(m)} K^2 + d^{(n)} \left| \sum_{k=0}^{K-1} e^{i(k(\Delta\omega)\tau + (\phi_k^{(n)} - \phi_k^{(m)}))} \right|^2 +$$

$$2 d^{(m)} d^{(n)} K \Re \left(e^{-i\omega_0 \tau} \sum_{k=0}^{K-1} e^{-i(k(\Delta\omega)\tau + (\phi_k^{(n)} - \phi_k^{(m)}))} \right) \tag{8}$$

When $d^{(n)} = 0$ there is no interference and we are back to the single-user case. hence we assume $d^{(n)} = 1$ from here on.

Note 2. In the synchronous $\tau = 0$ case, if $\Phi^{(m)}$ and $\Phi^{(n)}$ are Walsh-Hadamard sequences, (i.e., each $\{\exp(i\phi_k^{(\ell)})\}$ is a sequence in the Walsh-Hadamard sequence family), the two summations in (8) become zero, and there is no interference. This is of course clearly not the case when $\tau \neq 0$ (see Example 2).

Setting

$$\Theta_{nm}(\tau) = \sum_{k=0}^{K-1} e^{-i[k(\Delta\omega)\tau + (\phi_k^{(n)} - \phi_k^{(m)})]} \tag{9}$$

and noting that

$$|\, \mathfrak{Re}\left(e^{-i\omega_0\tau}\Theta_{nm}(\tau)\right)| \leq |\Theta_{nm}(\tau)| \tag{10}$$

we obtain the following upper and lower bounds for Equation (8):

$$d^{(m)}K^2 + |\Theta_{nm}(\tau)|^2 - 2d^{(m)}K\,|\Theta_{nm}(\tau)| \leq A$$
$$\leq d^{(m)}K^2 + |\Theta_{nm}(\tau)|^2 + 2d^{(m)}K\,|\Theta_{nm}(\tau)| \tag{11}$$

It follows that minimization of $|\Theta_{nm}(\tau)|$ for all τ is a reasonable criterion for signal design.

Note 3. The above generalizes in straightforward fashion to the case of more than 2 users.

3 Connection with PAPR Problem

Our objective is thus the design of sequences of length K such that:

$$\max_{\tau} |\Theta_{nm}(\tau)| = \max_{\tau\in[0,\frac{2\pi}{\Delta\omega})} \left| \sum_{k=0}^{K-1} e^{-i[k(\Delta\omega)\tau + (\phi_k^{(n)} - \phi_k^{(m)})]} \right| \tag{12}$$

is minimized for every sequence pair $\{\phi_k^{(n)}\}, \{\phi_k^{(m)}\}$. Equivalently, we seek to minimize

$$\max_{\tau\in[0,1)} \left| \sum_{k=0}^{K-1} e^{-ik2\pi\tau} e^{-i(\phi_k^{(n)} - \phi_k^{(m)})} \right| \tag{13}$$

The design of sequences with minimum PAPR (peak to average power ratio) crops up in conjunction with signal design for OFDM systems [17,18,19,20]. Since designing for low PAPR is hard, the common design approach is to design for low PMEPR (peak-to-mean envelope power ratio) which is more tractable. The PEMPR problem (see [20]) is one of designing sequences $\{a_k\}$ which minimize :

$$\max_{\tau\in[0,1)} \left| \frac{1}{K} \sum_{k=0}^{K-1} a_k e^{-ik2\pi\tau} \right|^2 \tag{14}$$

As can be seen, in our problem we are interested in phase sequences $\Phi^{(m)}$ and $\Phi^{(n)}$ such that $\exp(-i(\Phi^{(n)} - \Phi^{(m)}))$ is a sequence with good PMEPR.

The results in [20] as applied to the present situation are stated below. Let

$$M_d^{(K)} = \max_{j=0,\cdots,K-1} \left| \sum_{k=0}^{K-1} e^{-ik2\pi\left(\frac{j}{K}\right)} e^{-i(\phi_k^{(n)}-\phi_k^{(m)})} \right| \tag{15}$$

and

$$M_c^{(K)} = \max_{\tau\in[0,1)} \left| \sum_{k=0}^{K-1} e^{-ik2\pi\tau} e^{-i(\phi_k^{(n)}-\phi_k^{(m)})} \right| \tag{16}$$

From [20], we have that

Proposition 1. $M_d^{(K)} \geq \sqrt{K}$.

Proposition 2. *For $K > 3$:*

$$\frac{M_c^{(K)}}{M_d^{(K)}} < \frac{2}{\pi} \ln K + 1.132 + \frac{3}{K} \tag{17}$$

The above proposition is a special form of the general bound derived in [20]:

Proposition 3. *For $K > 3$:*

$$\frac{2}{\pi} \ln K + 0.603 - \frac{1}{6K} < \max_{F_K(t)} \left\{ \frac{M_c(F_K)}{M_d(F_K)} \right\} < \frac{2}{\pi} \ln K + 1.132 + \frac{3}{K} \tag{18}$$

in which:

$$F_K(t) = \sum_{k=0}^{K-1} a_k e^{2\pi ikt}; \quad \text{Such that} \quad \sum_{k=0}^{K-1} |a_k|^2 = K \tag{19}$$

and

$$M_d(F_K) = \max_{j=0,\cdots,K-1} \left| F_K\left(\frac{j}{K}\right) \right| \quad , \quad M_c(F_K) = \max_{t\in[0,1)} |F_K(t)|$$

The implication of Proposition 2 is that, if we design sequences with good asynchronous properties for sufficiently many samples of τ, it is guaranteed that the same sequences have good asynchronous properties for all values of τ.

4 Construction of Good Asynchronous Sequences for Phase Encoding OCDMA

In this section we use generalized bent functions to design sequences with good asynchronous properties. Some preliminaries on generalized bent functions that we will use are introduced in following:

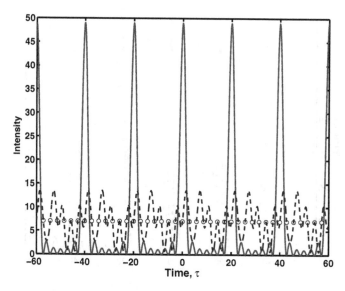

Fig. 2. An Example of Application of Theorem 6 with $K = 7$

4.1 Generalized Bent Functions

Definition 1. *[21] Let J_q^m denote the set of m-tuples with elements drawn from the set of integers modulo q, $w = e^{i\left(\frac{2\pi}{q}\right)}$ and g a complex-valued function defined on J_q^m. The Fourier transform of g is then defined to be the function G given by:*

$$G(\lambda) = \frac{1}{\sqrt{q^m}} \sum_{x \in J_q^m} g(x) w^{-\lambda^t x}, \quad \lambda \in J_q^m \tag{20}$$

Definition 2. *[21] A function f, $f : J_q^m \to J_q^1$ is said to be bent if the Fourier transform coefficients of w^f all have unit magnitude.*

Proposition 4. *[21]Every affine or linear translate of a bent function is also bent.*

Proposition 5. *[21]Let q be odd. Then the function f over J_q^1 defined by:*

$$f(k) = k^2 + ck + d \quad all \quad k \in J_q^1 . \tag{21}$$

is bent for all $c \in J_q^1$ and $d \in J_q^1$.

4.2 Construction

Theorem 6. *Let*

$$\Phi = \{\Phi^{(\ell)} \mid \Phi^{(\ell)} = \{\phi_0^\ell, \phi_1^\ell, \cdots, \phi_{K-1}^\ell\}, \ell \in \{1, 2, \cdots, L\}\}$$

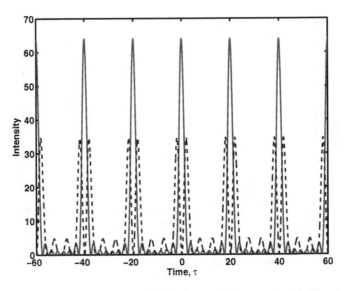

Fig. 3. An Example of Walsh-Hadamard Sequences with $K = 8$

be a family of phase sequences such that the difference sequence is associated to a bent function, i.e.,

$$\Phi^{(n)} - \Phi^{(m)} = \frac{2\pi}{K}(f(0), f(1), \cdots, f(K-1)), n \neq m$$

where $f(x)$ is a bent function over J_K. Then $\max |\Theta_{nm}(\tau)|$ is as small as can possibly be over multiples τ of $\frac{2\pi}{K(\Delta w)}$, and thus these phase sequences are suitable for use in asynchronous phase-encoded OCDMA systems.

In particular, the phase sequences

$$\phi_k^{(m)} = (k^3 + a_m k^2 + b_m k + c_m)\frac{2\pi}{K},$$

$$a_m, b_m, c_m \in \mathbb{Z}_K; \; m \neq n : a_m \neq a_n; \; K \; a \; prime > 2$$

are suitable for use as for asynchronous phase encoding OCDMA systems with K modes, where K is an odd prime.

Example 1. Figure 2 shows the application of the construction of Theorem 6, where $K = 7$, $w_0 = \frac{\pi}{4}$, $\Delta w = \frac{\pi}{10}$, $(a_m, b_m, c_m) = (2, 5, 3)$ and $(a_n, b_n, c_n) = (5, 4, 1)$. For this system $\Phi^{(m)} = (\frac{6\pi}{7}, \frac{8\pi}{7}, \frac{2\pi}{7}, 0, 0, 0, \frac{12\pi}{7})$ and $\Phi^{(n)} = (\frac{2\pi}{7}, \frac{8\pi}{7}, \frac{4\pi}{7}, \frac{2\pi}{7}, 0, \frac{10\pi}{7}, \frac{2\pi}{7})$. In this figure the solid line is the output of MLL as seen after the photo detector and the dashed line shows $|\Theta_{nm}(\tau)|^2$ at the output. Here the circles are samples of $|\Theta_{nm}(\tau)|^2$ for $\tau = \frac{2\pi j}{K(\Delta w)}$. As can be seen, all these values are equal to $K = 7$. It can be observed that $|\Theta_{nm}(\tau)|^2$ is low for all values of τ and the phase sequences are thus applicable for asynchronous transmission.

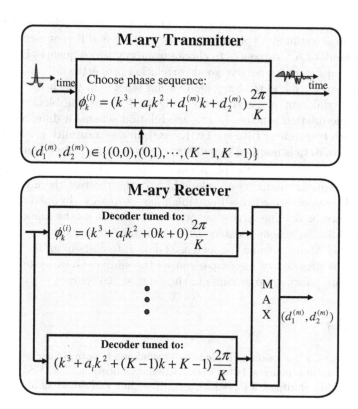

Fig. 4. Block Diagram of M-ary Modulation Transmitter and Receiver

Example 2. Figure 3 shows the application of Walsh-Hadamard sequences for asynchronous systems. In this graph $K = 8$, $\omega_0 = \frac{\pi}{4}$, $\Delta\omega = \frac{\pi}{10}$, $\Phi^{(m)} = (0,0,0,0,0,0,0,0)$ and $\Phi^{(n)} = (\pi,\pi,\pi,\pi,0,0,0,0)$. In this figure the solid line is the output of MLL as it is seen after photo detector and the dashed line is showing $|\Theta_{nm}(\tau)|^2$ at the output. As can be seen the system has no interference for $\tau = 0$ (synchronous case) while with a little deviation from $\tau = 0$, $|\Theta_{nm}(\tau)|^2$ increases dramatically. Because of the high peak of $|\Theta_{nm}(\tau)|^2$, these phase sequences are not suitable for asynchronous transmission.

5   M-ary Modulation

Traditionally OOK modulation is used with phase encoding OCDMA. The spectral efficiency of the system can be increased if we can move from binary modulation to M-ary modulation. As we will see the sequences introduced in Theorem 6 can be used for M-ary modulation without increasing the amount of interference.

In the phase sequences introduced in Theorem 6, two different users should have sequences with different a_ℓ(square term coefficient), while when we fix a_ℓ

we can choose b_ℓ and c_ℓ(linear coefficient and constant term) arbitrarily. In other words we can accommodate phase sequences only for K different users, but each user has a bank of K^2 sequences to choose from(any two sequences chosen from banks of two different users have good correlation properties).

Now we introduce a new M-ary modulation scheme in which the data is encoded in the different phase sequences. Figure 4 shows the block diagram of the M-ary modulation scheme. In this modulation scheme a different a_ℓ is assigned to each transmitter (like the OOK case). In the transmitter, each block of $2 \log_2 K$ bits of data is mapped to a different vector (b_ℓ, c_ℓ). Based on the (b_ℓ, c_ℓ) vector, the transmitter uses the phase encoder $\phi_k^{(\ell)} = (k^3, a_\ell k^2 + b_\ell k + c_\ell)\frac{2\pi}{K}$ out of its phase sequence bank to encode MLL. At the receiver there is a bank of K^2 parallel decoders tuned to all possible phase sequences in one transmitter's phase sequence bank. The largest output will be chosen as the input symbol.

This modulation scheme enables us to transmit more data per symbol. As with any other M-ary scheme the increased data rate costs in an increase in the system's error probability in comparison to the similar OOK system. That is because we are packing more points in the same signal space.

References

1. Salehi, J.A.: Code division multiple-access techniques in optical fiber networks–part I: Fundamental principles. IEEE Trans. Commununication 37, 824–833 (1989)
2. Weiner, A.M., Heritage, J.P., Salehi, J.A.: Encoding and decoding of femtosecond pulses. Opt. Lett. 13, 300–302 (1988)
3. Salehi, J.A., Weiner, A.M., Heritage, J.P.: Coherent ultrashort light pulse code-division multiple access commuincation systems. IEEE Journal of Lightwave Tech. 8, 478–491 (1990)
4. Chung, F., Salehi, J.A., Wei, V.K.: Optical orthogonal codes: Design, analysis, and applications. IEEE Trans. Information Theory 35, 595–604 (1989)
5. Chung, H., Kumar, P.V.: Optical orthogonal codes – new bounds and an optimal construction. IEEE Trans. Information Theory 36, 866–873 (1990)
6. Nguyen, Q.A., Györfi, L., Massey, J.L.: Constructions of binary constant-weight cyclic codes and cyclically permutable codes. IEEE Trans. Information Theory 38, 940–949 (1992)
7. Moreno, O., Zhang, Z., Kumar, P.V., Zinoviev, V.: New constructions of optimal cyclically permutable constant weight codes. IEEE Trans. Information Theory 41, 448–455 (1995)
8. Buratti, M.: A powerful method for constructing difference families and optimal optical orthogonal codes. Designs, Codes and Cryptography 5, 13–25 (1995)
9. Ding, C., Xing, C.: Several classes of $(2^m - 1, w, 2)$ optical orthogonal codes. Discrete Applied Mathematics 128, 103–120 (2003)
10. Miyamoto, N., Mizuno, H., Shinohara, S.: Optical orthogonal codes obtained from conics on finite projective planes. Finite Fields and Their Applications 10, 405–411 (2004)
11. Moreno, O., Omrani, R., Kumar, P.V., Lu, H.: A generalized Bose-Chowla family of optical orthogonal codes and distinct difference sets. IEEE Trans. Information Theory 53, 1907–1910 (2007)

12. Moreno, O., Omrani, R., Kumar, P.V.: New bounds on the size of optical orthogonal codes, and constructions. The IEEE Transactions on Information Theory (to be submitted)
13. Kavehrad, M., Zaccarin, D.: Optical code-division-multiplexed systems based on spectral encoding of noncoherent sources. IEEE Journal of Lightwave Tech. 13, 534–545 (1995)
14. Galli, S., Menendez, R., Toliver, P., Banwell, T., Jackel, J., Young, J., Etemad, S.: DWDM compatible spectrally phase encoded optical CDMA. In: Proc. Globecom Conf., pp. 1888–1894 (2004)
15. Hernandez, V.J., Du, Y., Cong, W., Scott, R.P., Li, K., Heritage, J.P., Ding, Z., Kolner, B.H., Yoo, S.J.B.: Spectral phase-encoded time-spreading (SPECTS) optical code-division multiple-access for terabit optical access networks. IEEE Journal of Lightwave Tech. 22, 2671–2679 (2004)
16. Etemad, S., Toliver, P., Menendez, R., Young, J., Banwell, T., Galli, S., Jackel, J., Delyett, P., Price, C., Turpin, T.: Spectrally efficient optical CDMA using coherent phase-frequency coding. IEEE Photon. Technol. Lett. 17, 929–931 (2005)
17. Davis, J.A., Jedwab, J.: Peak-to-mean power control in OFDM, golay complementary sequences, and Reed-Muller codes. IEEE Trans. Information Theory 465, 2397–2417 (1999)
18. Paterson, K.G., Tarokh, V.: On the existence and construction of good codes with low peak-to-average power ratios. IEEE Trans. Information Theory 46, 1974–1987 (2000)
19. Paterson, K.G.: Sequences for OFDM and multi-code CDMA: Two problems in algebraic coding theory. In: Helleseth, T., Kumar, P.V., Yang, K. (eds.) SETA 2001, pp. 46–71. Springer, Heidelberg (2002)
20. Litsyn, S., Yudin, A.: Discrete and continuous maxima in multicarrier communication. IEEE Trans. Information Theory 51, 919–928 (2005)
21. Kumar, P.V., Scholtz, R.A., Welch, L.R.: Generalized bent functions and their properties. Journal of Combinatorial Theory. Series A 40, 90–107 (1985)
22. Stapleton, A., Shafiiha, R., Akhavan, H., Farrell, S., Peng, Z., Choi, S.J., Marshal, W., O'brien, J.D., Dapkus, P.D.: Experimental measurement of optical phase in microdisk resonators. In: IEEE/LEOS Summer Topical Meetings, pp. 54–57 (June 2004)
23. Stapleton, A., Farrell, S., Akhavan, H., Shafiiha, R., Peng, Z., Choi, S.J., Marshal, W., O'brien, J.D., Dapkus, P.D.: Optical phase characterization of active semiconductor microdisk resonators in transmission. Applied Physics Letters 88 (January 2006)

Packing Centrosymmetric Patterns of n Nonattacking Queens on an $n \times n$ Board

Herbert Taylor

1101 Loma Vista Court
South Pasadena, CA 91030, USA

Abstract

n	1	2	3	4	5	6	7	8	9	10	11
$SQ(n)$	1	0	0	2	1	4	1	4	1	8	1
$Q(n)$	1	0	0	2	5	4	7	6	?	?	11

n	12	13	14	15	16	17	18	19	20	21	22
$SQ(n)$	10	1	12	1	14	1	16	1	?	1	20
$Q(n)$	?	13	?	?	?	17	?	19	?	?	?

$SQ(n)$ is the maximum number of patterns that can sit on the $n \times n$ board where each pattern consists of n nonattacking Queens placed symmetrically around the center. Each square of the board has at most one Queen. $Q(n)$ is the same except that "placed symmetrically around the center" is not required.

Dedicated to Solomon W. Golomb on his 75th birthday

1 Introduction

Here we have two sequences of packing problems where the things to be packed are solutions to the n queens problem.

$SQ(n)$ is the maximum number of patterns that can sit simultaneously on the $n \times n$ board, where each pattern consists of n nonattacking Queens placed symmetrically around the center. Each square of the board has at most one Queen. $Q(n)$ is the same except that "placed symmetrically around the center" is not required.

The original n queens problem around 1850 was to place 8 queens on the chessboard so that no two attacked each other. See [2]. Later, when many solutions had been found, the problem became to count the solutions. Now the Encyclopedia of Integer Sequences published in 1995, by N.J.A. Sloane and Simon Plouffe gives the number of solutions up to $n = 20$, and the number of symmetric solutions up to $n = 19$.

Table 1 shows what we know about n queens, $Q(n)$, and $SQ(n)$.

We may need the following definition: a latin square is *vatican* if for any two symbols a and b and for any k, there is at most one row in which a sits k steps to the right of b.

The only Vatican square known to exist are given by the multiplication table modulo p for a prime number p.

S.W. Golomb et al. (Eds.): SSC 2007, LNCS 4893, pp. 106–118, 2007.

Table 1. Status of n queens problem, $Q(n)$, and $SQ(n)$

n	# patterns	$Q(n)$	# symm	$SQ(n)$
1	1	1	1	1
2	0	0	0	0
3	0	0	0	0
4	2	2	2	2
5	10	5	2	1
6	4	4	4	4
7	40	7	8	1
8	92	6	4	4
9	352	$7 \leq ? \leq 9$	16	1
10	724	$8 \leq ? \leq 10$	12	8
11	2680	11	48	1
12	14200	$10 \leq ? \leq 12$	80	10
13	73712	13	136	1
14	365596	$12 \leq ? \leq 14$	420	12
15	2279184	?	1240	1
16	14772512	$14 \leq ? \leq 16$	2872	14
17	95815104	17	7652	1
18	666090624	$16 \leq ? \leq 18$	18104	16
19	4968057848	19	50184	1
20	39029188884	$? \leq 20$		$? \leq 18$
21		$? \leq 21$		1
22		$20 \leq ? \leq 22$		20

2   Results

This paper contains four cases. These are **CASE I:** $SQ(n)$ when n is odd; **CASE II:** $SQ(n)$ when n is even; **CASE III:** $Q(n)$ when n is odd; and **CASE IV:** $Q(n)$ when n is even.

In any case any diagonal can have at most one Queen from one pattern of n nonattacking Queens.

CASE I: $SQ(n)$ when n is odd

The center row and center column meet on the center square, which therefore must have a Queen from any pattern which is both odd and centrosymmetric.

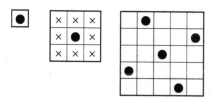

Fig. 1. $SQ(1) = 1$, $SQ(3) = 0$, and $SQ(5) = 1$

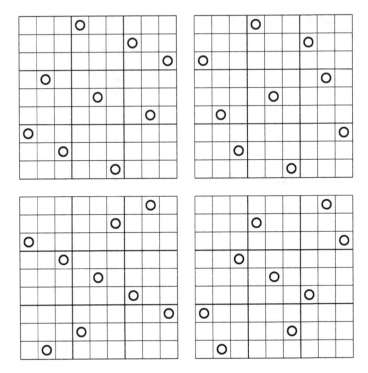

Fig. 2. $SQ(9) = 1$

Figure 1 shows that $SQ(1) = 1$, $SQ(3) = 0$ and $SQ(5) = 1$. Each of the four figures in Fig. 2 shows $SQ(9) = 1$. First conjecture is the following:

Conjecture 1. $SQ(n) = 1$ for odd $n > 3$.

CASE II: $SQ(n)$ when n is even

In this case, the center is not a square of the board so both main diagonals must be empty. Thus we know that when n is even, $SQ(n) \leq n - 2$.

Theorem 2. If $n + 1$ is an odd prime, then $SQ(n) = n - 2$.

Proof. Write the multiplication table for the prime $n + 1$, putting the first column $1, 2, 3, ..., n$ and the first row $1^{-1}, 2^{-1}, 3^{-1}, ..., n^{-1}$. So, one main diagonal will have all 1's, and the other diagonal will have all n's. Golomb's observation that this is a Vatican square leads to the observation that on any diagonal other than the two main diagonals all the symbols must be different. In other words each of the symbols $2, 3, ..., n - 1$ forms a pattern of n nonattacking Queens on an $n \times n$ board. Centrosymmetry holds because in a prime field we have $-(a^{-1}) = (-a)^{-1}$. $\square$

1	6	4	3	9	2	8	7	5	10
2	1	8	6	7	4	5	3	10	9
3	7	1	9	5	6	2	10	4	8
4	2	5	1	3	8	10	6	9	7
5	8	9	4	1	10	7	2	3	6
6	3	2	7	10	1	4	9	8	5
7	9	6	10	8	3	1	5	2	4
8	4	10	2	6	5	9	1	7	3
9	10	3	5	4	7	6	8	1	2
10	5	7	8	2	9	3	4	6	1

Fig. 3. $SQ(10) = 8$

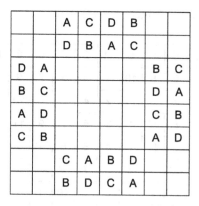

Fig. 4. $SQ(8) = 4$

Another proof adapted from Golomb's construction of "superqueens" [3] pictures each pattern as a line through the origin on a $p \times p$ vector space over $GF(p)$.

Yet another proof starts with the Welch construction for Costas arrays as in [7].

All three proofs give us essentially the same picture. See Fig. 3

What about $SQ(n)$ when $n+1$ is odd but not prime? From [8] we know that only four centrosymmetric patterns of eight nonattacking Queens exist on the 8×8 board. See Fig. 4.

A rash conjecture was made in [7] that if $SQ(n) = n - 2$ then $n + 1$ must be prime. See Fig. 5 for a counterexample. Now we are left with the following open question:

Question 3. $SQ(n) =$? when $n + 1 > 15$ is odd but not prime.

	D	f	g	A	w	m	z	x	C	e	h	B	
A		m	x	f	e	D	B	g	h	w	z		C
e	x		C	z	B	h	f	D	m	A		w	g
h	m	B		g	A	x	w	C	e		D	z	f
D	e	w	h		z	C	A	m		f	x	g	B
z	f	C	D	w		e	g		x	B	A	h	m
x	A	g	m	B	f			h	D	z	e	C	w
w	C	e	z	D	h			f	B	m	g	A	x
m	h	A	B	x		g	e		w	D	C	f	z
B	g	x	f		m	A	C	z		h	w	e	D
f	z	D		e	C	w	x	A	g		B	m	h
g	w		A	m	D	f	h	B	z	C		x	e
C		z	w	h	g	B	D	e	f	x	m		A
	B	h	e	C	x	z	m	w	A	g	f	D	

Fig. 5. $SQ(14) = 12$

1	3	5	7	9	11	2	4	6	8	10
2	4	6	8	10	1	3	5	7	9	11
3	5	7	9	11	2	4	6	8	10	1
4	6	8	10	1	3	5	7	9	11	2
5	7	9	11	2	4	6	8	10	1	3
6	8	10	1	3	5	7	9	11	2	4
7	9	11	2	4	6	8	10	1	3	5
8	10	1	3	5	7	9	11	2	4	6
9	11	2	4	6	8	10	1	3	5	7
10	1	3	5	7	9	11	2	4	6	8
11	2	4	6	8	10	1	3	5	7	9

Fig. 6. $Q(11) = 11$

CASE III: $Q(n)$ when n is odd

Theorem 4. If $n = 6m + 1$ or $n = 6m - 1$, then $Q(n) = n$.

Proof. An easy construction is as follows. Start with an $n \times n$ board. In the first row write $1, 3, 5, ..., n, 2, 4, 6, ..., n - 1$. In row $k + 1$ add 1 to each number in row k mod n. The resulting latin square, as it stands, shows that $Q(n) = n$, if $n = 6m + 1$ or $n = 6m - 1$. See Fig. 6. $\square$

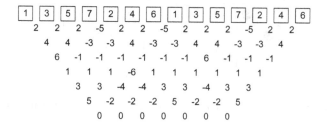

```
  2   2   2  -5   2   2  -5   2   2   2  -5   2   2
    4   4  -3  -3   4  -3  -3   4   4  -3  -3   4
      6  -1  -1  -1  -1  -1  -1   6  -1  -1  -1
        1   1   1  -6   1   1   1   1   1   1
          3   3  -4  -4   3   3  -4   3   3
            5  -2  -2  -2   5  -2  -2   5
              0   0   0   0   0   0   0
```

Fig. 7. $Q(7) = 7$

1	B	3		d	4	m		2
	4	2	m	B	3		1	d
m			d	1	2	B	3	4
3	2	1	B	4	d			m
B	m	4	3		1	2	d	
	d	B		2	m	3	4	1
2	1		4	3		d	m	B
d	3	m	1		B	4	2	
4		d	2	m		1	B	3

Fig. 8. $Q(9) \geq 7$

I believe this theorem is also proved in [8].

To check that each row gives a pattern of n nonattacking Queens in Gauss's Arithmetization we can use a difference triangle extended far enough to include all the shifts of the pattern toroidal fashion. See Fig. 7.

When $n = 6m + 3$ the situation is baffling. $Q(3) = 0$ is the only $n = 6m + 3$ for which I know the exact value of $Q(n)$. Any comments will be very welcome. The present state of affairs for $n = 9$ is that $7 \leq Q(9) \leq 9$. See Fig. 8. See "A Simple Game."

A Simple Game

The board can be 9×9. Player A has nine pieces each labeled A. Player B has nine pieces each labeled B. The players take turns putting one of their letters on the board. The letter can go on any unoccupied square, but never allowing two of the same letter in any row, column, or diagonal. If all nine A's and all nine B's get put on the board, the game is a draw. Otherwise the first player who cannot put another letter is the loser.

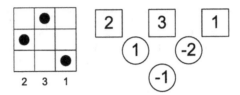

Fig. 9. A 3 × 3 pattern that fails to be a nonattacking Queens pattern

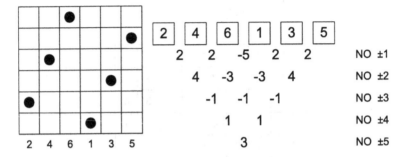

Fig. 10. A 6 × 6 pattern that is a nonattacking Queens pattern

3	A	4		D	1		2
	2	D	1	4	A	3	
A	4		3	2		1	D
1		2	D		3	A	4
2		1	A		4	D	3
D	3		4	1		2	A
	1	A	2	3	D	4	
4	D	3		A	2		1

Fig. 11. $Q(8) \geq 6$

To avoid draws the game could continue with nine each of C, D, E, F, G, and H. If the players reach a draw in that extended game it will mean $Q(9) \geq 8$, which will be worth a prize.

Gauss' Arithmetization and Difference Triangle

Gauss' idea was to represent an $n \times n$ permutation matrix as a permutation of the numbers from 1 to n so that the number in the j-th position is the height of the dot in the j-th column. For example,

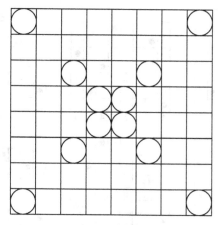

Fig. 12. A configuration of twelve squares on 8×8 board that shows $Q(8) < 8$

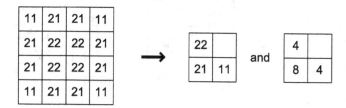

Fig. 13. A 4×4 board and its symmetry coloring

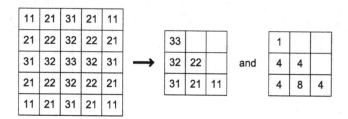

Fig. 14. A 5×5 board and its symmetry coloring

The test for nonattacking Queens is that for every k from 1 to $n-1$, $(j+k)$-th height minus j-th height shall never be $+k$ nor $-k$. To see at a glance whether or not a permutation passes Gauss' test, picture it in a difference triangle as shown in Figures 9 and 10.

CASE IV: $Q(n)$ when n is even

When $n > 8$ we don't have a proof that $Q(n) < n$.

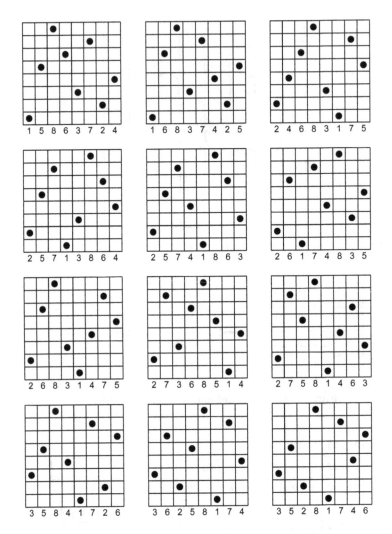

Fig. 15. All the distinct twelve nonattacking Queens patterns for $n = 8$

44			
43	33		
42	32	22	
41	31	21	11

4			
8	4		
8	8	4	
8	8	8	4

10 colors for the 8×8   #squares/color

Fig. 16. Symmetry coloring of an 8×8 board

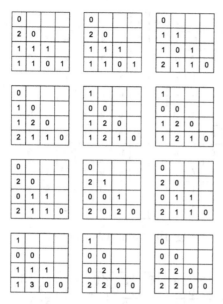

Fig. 17. Number of dots in symmetry coloring, corresponding to Figure 15

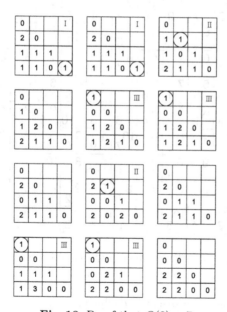

Fig. 18. Proof that $Q(8) < 7$

Question 5. Do there exist any even n for which $Q(n) > n - 2$?

Now we will conclude with an explanation of the proof that $Q(8) = 6$. By trial and error the picture of Fig. 11 was discovered. Then a configuration of twelve

9	0	2	4	6	8	10	1	3	5	7
3	5	7	9	0	2	4	6	8	10	1
8	10	1	3	5	7	9	0	2	4	6
2	4	6	8	10	1	3	5	7	9	0
7	9	0	2	4	6	8	10	1	3	5
1	3	5	7	9	0	2	4	6	8	10
6	8	10	1	3	5	7	9	0	2	4
0	2	4	6	8	10	1	3	5	7	9
5	7	9	0	2	4	6	8	10	1	3
10	1	3	5	7	9	0	2	4	6	8
4	6	8	10	1	3	5	7	9	0	2

Fig. 19. 3-dimensional $Q(11)$

9_1	0_3	2_5	4_7	6_9	8_0	10_2	1_4	3_6	5_8	7_{10}
3_2	5_4	7_6	9_8	0_{10}	2_1	4_3	6_5	8_7	10_9	1_0
8_3	10_5	1_7	3_9	5_0	7_2	9_4	0_6	2_8	4_{10}	6_1
2_4	4_6	6_8	8_{10}	10_1	1_3	3_5	5_7	7_9	9_0	0_2
7_5	9_7	0_9	2_0	4_2	6_4	8_6	10_8	1_{10}	3_1	5_3
1_6	3_8	5_{10}	7_1	9_3	0_5	2_7	4_9	6_0	8_2	10_4
6_7	8_9	10_0	1_2	3_4	5_6	7_8	9_{10}	0_1	2_3	4_5
0_8	2_{10}	4_1	6_3	8_5	10_7	1_9	3_0	5_2	7_4	9_6
5_9	7_0	9_2	0_4	2_6	4_8	6_{10}	8_1	10_3	1_5	3_7
10_{10}	1_1	3_3	5_5	7_7	9_9	0_0	2_2	4_4	6_6	8_8
4_0	6_2	8_4	10_6	1_8	3_{10}	5_1	7_3	9_5	0_7	2_9

Fig. 20. Orthogonality of $Q(11)$ and $3DQ(11)$

squares on the 8×8 board was discovered which has the property that any pattern of 8 nonattacking Queens can take at most one square in the configuration. See Fig. 12. Any pattern of 8 nonattacking Queens can sit on at most one square of the configuration. Thus 8 patterns would leave four squares of the configuration empty, and consequently, $Q(8) < 8$. It is now sufficient to show that $Q(8) < 7$. For this, we need the concept of symmetry coloring.

Symmetry Coloring

We can color the squares of the $n \times n$ board by symmetry, that is, two squares get the same color if one can be moved to the other by rotations and/or reflections of the board.

For example, the 4×4 board gets three colors 11, 22, and 21, as shown in the first of Fig. 13. Here, the color 11 is on 4 squares, the color 22 is on 4 squares, and the color 21 is on 8 squares. We can eightfold this so that it looks like that in the middle of Fig. 13, and it can be represented as shown in the third of Fig. 13. Another example on the 5×5 board is shown in Fig. 14.

To show that $Q(8) < 7$, we first found that there exist exactly 12 distinct 8×8 nonattacking Queens patterns [8] with respect to the symmetry of all rotations and reflections of the board. They are shown in Fig. 15.

By symmetry coloring, we distinguish the colors of each square of an 8×8 board as $11, 22, 33, 44$ appearing 4 times, and $21, 31, 41, 32, 42, 43$ appearing 8 times each, respectively, as shown in Fig. 16.

Now, each pattern of 8×8 nonattacking Queens patterns in Fig. 15 can be represented as shown in Fig. 17. Here, the number in each square is the number of dots of each pattern with respect to the symmetry of squares.

The twelve patterns are now classified into I, II, and III, as shown in Fig. 18, and we are now ready to show that it is impossible to place any 7 of them into a single 8×8 board. To put seven patterns there cannot be two empty squares in any row or column. Therefore, we must take two patterns in I to satisfy color 11. Otherwise, see that there exist two empty squares in a row or a column passing through the corner squares. We must take two more in II for color 33, and take two more in III for color 44. But then, with six patterns, color 41 is **full** with eight. A seventh pattern would overfill color 41. This completes the proof.

Afterword

During the talk, Sol Golomb asked for a renumbering of the figure $Q(11) = 11$ to make an example of n^2 nonattacking 3D Queens on a board which is $n \times n \times n$.

As presented in [4], any latin square can be seen as a figure composed of n^2 nonattacking 3D Rooks on a board which is $n \times n \times n$. Here we have Fig. 19 obtained by permuting rows of Fig. 6. This construction will surely work when n is prime and $n > 7$. But a question remains open for $n = 25$ or $n = 35$ or in general when $n = 6m + 1$ or $n = 6m - 1$ but n is not prime.

Another question which will necessitate much further study appeared unexpectedly with the discovery that "$Q(11) = 11$" and "3DQ(11)" shown in Fig. 6 and Fig. 19 form a pair of orthogonal latin squares. See Fig. 20.

References

1. Ginsburg, J.: Gauss's arithmetization of the problem on n queens. Scripta math. 5, 63–66 (1939)
2. Campbell, P.J.: Gauss and the 8-queens problem: A study in miniature of the propagation of historical error. Historia mathematica 4, 397–404 (1977)

3. Golomb, S.W.: Sphere packing, coding metrics, and chess puzzles. In: Proceedings of the second Chapel Hill Conference on Combinatorial Mathematics and its Applications, U. No. Carolina at Chapel Hill, may (1970)
4. Golomb, S.W., Posner, E.C.: Rook domains, latin squares, affine planes, and errordistributing codes. IEEE Trans. Inform. Th. 10-3, 196–208 (1964)
5. Golomb, S.W., Taylor, H.: Constructions and properties of Costas arrays. IEEE Proc. 72, 1143–1163 (1984)
6. Etzion, T.: Combinatorial designs derived from Costas arrays. In: Sequences, Proceedings, Salerno, Italy (June 1989)
7. Taylor, H.: Singly periodic Costas arrays are equivalent to polygonal path Vatican squares. In: Mathematcial properties of sequences and other combinatorial structures, pp. 45–53. Kluwer Academic Publishers, Dordrecht (2003)
8. Kraitchik, M.: Mathematical Recreation (1944)
9. Golomb, S.W.: A mathematical theory of discrete classification. In: Proceedings of fourth London Symosium on Imformation Theory, pp. 404-425 (1959)

Cyclotomic Mapping Permutation Polynomials over Finite Fields

Qiang Wang[*]

School of Mathematics and Statistics, Carleton University,
1125 Colonel By Drive, Ottawa, Ontario, K1S 5B6, Canada,
wang@math.carleton.ca,
http://math.carleton.ca/~wang.html

Abstract. We explore a connection between permutation polynomials of the form $x^r f(x^{(q-1)/l})$ and cyclotomic mapping permutation polynomials over finite fields. As an application, we characterize a class of permutation binomials in terms of generalized Lucas sequences.

Keywords: permutation polynomials, cyclotomic mappings, generalized Lucas sequences, Dickson polynomials, finite fields.

1 Introduction

Let p be prime and $q = p^m$. A polynomial is a permutation polynomial (PP) of a finite field $\mathbb{F}_q$ if it induces a bijective map from $\mathbb{F}_q$ to itself. The study of permutation polynomials of a finite field goes back to 19-th century when Hermite and later Dickson pioneered this area of research. In recent years, interests in permutation polynomials have significantly increased because of their potential applications in public key cryptosystems ([12],[13],[14]), RC6 block ciphers ([21], [22]), combinatorial designs like de Bruijn sequences ([6]), Tuscan-k arrays ([8]), and Costas arrays ([5], [11]), among many others. Permutation polynomials are also used in coding theory, for instance, permutation codes in power communications ([7]), and interleavers in Turbo codes ([26]) etc. In some of these applications, the study of permutation polynomials over finite fields has also been extended to the study of permutation polynomials over finite rings and other algebraic structures. For more background material on permutation polynomials we refer to Chap. 7 of [18]. For a detailed survey of open questions and recent results see [9], [15], [16], and [19].

Every polynomial $P(x)$ over $\mathbb{F}_q$ such that $P(0) = 0$ has the form $x^r f(x^s)$ with $r > 0$ and some positive integer $s \mid q - 1$. Here we are interested in permutation behavior of polynomials $P(x) = x^r f(x^s)$ over finite field $\mathbb{F}_q$, where $f(x)$ is an arbitrary polynomial of degree $e > 0$, $0 < r < q - 1$, and $q - 1 = ls$ for some positive integers l and s. In Sect. 2, we introduce the notion of r-th order cyclotomic mappings $f^r_{A_0, A_1, \cdots, A_{l-1}}$ of index l and reveal a simple and very useful connection between polynomials of the form $x^r f(x^s)$ and so-called r-th order

---

[*] The research is partially supported by NSERC of Canada.

S.W. Golomb et al. (Eds.): SSC 2007, LNCS 4893, pp. 119–128, 2007.

cyclotomic mapping polynomials. That is, $P(x) = x^r f(x^s) = f^r_{A_0, A_1, \cdots, A_{l-1}}(x)$ where $A_i = f(\zeta^i)$ for $0 \leq i \leq l-1$ and ζ is a primitive l-th root of unity. This provides us an easier way to study polynomials of the form $x^r f(x^s)$. Indeed, many different criteria of when these polynomials are permutation polynomials are summarized in Theorem 1. Some new classes of permutation polynomials are given to demonstrate the potential applications of these criteria. In particular, in Sect. 3, we characterize permutation binomials of the form $P(x) = x^r(x^{es} + 1)$ over $\mathbb{F}_q$ in terms of the generalized Lucas sequences of order $\frac{l-1}{2}$ over $\mathbb{F}_p$ as a concrete application (Theorem 3). Earlier study in this direction can be found in [1], [2] and [3].

2   Cyclotomic Mapping Permutation Polynomials

Let γ be a primitive element of $\mathbb{F}_q$, $q - 1 = ls$ for some positive integers l and s, and the set of all nonzero l-th powers of $\mathbb{F}_q$ be $C_0 = \{\gamma^{lj} : j = 0, 1, \cdots, s-1\}$. Then C_0 is a subgroup of $\mathbb{F}_q^*$ of index l. The elements of the factor group $\mathbb{F}_q^*/C_0$ are the *cyclotomic cosets*

$$C_i := \gamma^i C_0, \quad i = 0, 1, \cdots, l-1.$$

For any integer $r > 0$ and any $A_0, A_1, \cdots, A_{l-1} \in \mathbb{F}_q$, we define an *r-th order cyclotomic mapping* $f^r_{A_0, A_1, \cdots, A_{l-1}}$ *of index l* from $\mathbb{F}_q$ to itself by $f^r_{A_0, A_1, \cdots, A_{l-1}}(0) = 0$ and

$$f^r_{A_0, A_1, \cdots, A_{l-1}}(x) = A_i x^r \quad \text{if } x \in C_i, \quad i = 0, 1, \cdots, l-1.$$

Moreover, $f^r_{A_0, A_1, \cdots, A_{l-1}}$ is called an *r-th order cyclotomic mapping of the least index l* if the mapping can not be written as a cyclotomic mapping of any smaller index. The polynomial $f^r_{A_0, A_1, \cdots, A_{l-1}}(x) \in \mathbb{F}_q[x]$ of degree at most $q - 1$ representing the cyclotomic mapping $f^r_{A_0, A_1, \cdots, A_{l-1}}$ is called an *r-th order cyclotomic mapping polynomial*. In particular, when $r = 1$, it is known as a *cyclotomic mapping polynomial* (see [10] or [20]).

Let $\zeta = \gamma^s$ be a primitive l-th root of unity. Next we show that polynomials of the form $x^r f(x^s)$ and the r-th order cyclotomic mapping polynomials $f^r_{A_0, A_1, \cdots, A_{l-1}}(x)$ where $A_i = f(\zeta^i)$ for $0 \leq i \leq l-1$ are the same.

Lemma 1. *For any $r > 0$, $x^r f(x^s) = f^r_{A_0, A_1, \cdots, A_{l-1}}(x)$ where $A_i = f(\zeta^i)$ for $0 \leq i \leq l-1$.*

Proof. For any $x \in C_i$, $x = \gamma^{lj+i}$ for some $0 \leq j \leq s-1$. Hence $f(x^s) = f((\gamma^{lj}\gamma^i)^s) = f(\gamma^{is}) = f(\zeta^i) = A_i$. Therefore $P(x) = x^r f(x^s) = f^r_{A_0, A_1, \cdots, A_{l-1}}(x)$. $\square$

This simple connection provides us some useful criteria of when polynomials $P(x) = x^r f(x^s)$ are permutation polynomials of $\mathbb{F}_q$. It is well known that if $P(x) = x^r f(x^s)$ is a permutation polynomial of $\mathbb{F}_q$ then $(r, s) = 1$, where (r, s) denotes the greatest common divisor of r and s. Otherwise, let $(r, s) = d \neq 1$, then two distinct d-th roots of unity are mapped to the same element by $P(x)$, a contradiction. Moreover, we note that $A_j \neq 0$ for all $j = 0, \cdots l-1$ in any permutation polynomial $f^r_{A_0, A_1, \cdots, A_{l-1}}(x)$ since, otherwise, $f^r_{A_0, A_1, \cdots, A_{l-1}}(x)$ has more than 1 zeros. Hence we assume that $(r, s) = 1$ and $A_i \neq 0$ for $0 \leq i \leq l-1$ without loss of generality.

We first recall the following Lemma from [4].

Lemma 2. Let $l \mid q-1$ and $\xi_0, \xi_1, \cdots, \xi_{l-1}$ be some l-th roots of unity in $\mathbb{F}_q$. Then $\xi_0, \xi_1, \cdots, \xi_{l-1}$ are all distinct if and only if

$$\sum_{i=0}^{l-1} \xi_i^c = 0, \quad \text{for all } c = 1, \cdots, l-1.$$

Proof. For the sake of completeness, we include the proof from [4]. First note that for an l-th root of unity ξ, we have

$$1 + \xi + \cdots + \xi^{l-1} = \begin{cases} 0 \text{ if } \xi \neq 1, \\ l \text{ if } \xi = 1. \end{cases}$$

Now for $t = 0, \cdots, l-1$, let

$$h_t(x) = \sum_{j=0}^{l-1} \xi_t^{l-j} x^j.$$

We have

$$h_t(\xi_j) = \begin{cases} 0 \text{ if } t \neq j, \\ l \text{ if } t = j. \end{cases}$$

Let

$$h(x) = \sum_{t=0}^{l-1} h_t(x) = l + \sum_{j=1}^{l-1} \left(\sum_{t=0}^{l-1} \xi_t^{l-j} \right) x^j.$$

We consider h as a function from μ_l to $\mathbb{F}_q$. Since the degree of $h(x)$ is less than or equal to $l-1$, it is clear that $\xi_0, \xi_1, \cdots, \xi_{l-1}$ are all distinct if and only if $h(x) = l$. This implies the result. □

Theorem 1. Let p be prime and $q = p^m$, $q-1 = ls$ for some positive integers l and s, γ be a given primitive element of $\mathbb{F}_q$ and $\zeta = \gamma^s$ be a primitive l-th root of unity. Assume $P(x) = x^r f(x^s) = f^r_{A_0, A_1, \cdots, A_{l-1}}(x)$ with $(r, s) = 1$ and $A_i = f(\zeta^i) \neq 0$ for $0 \leq i \leq l-1$. Then the following are equivalent:

(a) $P(x) = x^r f(x^s)$ is a permutation polynomial of $\mathbb{F}_q$.
(b) $f^r_{A_0, A_1, \cdots, A_{l-1}}(x)$ is a permutation polynomial of $\mathbb{F}_q$.
(c) $A_i C_{ir} \neq A_{i'} C_{i'r}$ for any $0 \leq i < i' \leq l-1$.
(d) $Ind_\gamma(\frac{A_i}{A_{i'}}) \not\equiv r(i' - i) \pmod{l}$ for any $0 \leq i < i' \leq l-1$, where $ind_\gamma(a)$ is residue class $b \bmod q-1$ such that $a = \gamma^b$.
(e) $\{A_0, A_1\gamma^r, \cdots, A_{l-1}\gamma^{(l-1)r}\}$ is a system of distinct representatives of $\mathbb{F}_q^*/C_0$.

(f) $\{A_0^s, A_1^s\zeta^r, \cdots, A_{l-1}^s\zeta^{(l-1)r}\} = \mu_l$, *where μ_l is the set of all distinct l-th roots of unity.*

(g) $\displaystyle\sum_{i=0}^{l-1} \zeta^{cri} A_i^{cs} = 0$ *for all $c = 1, \cdots, l-1$.*

Proof. By Lemma 1, (a) and (b) are equivalent. Since $C_i = \{\gamma^{lj+i} : j = 0, 1, \cdots, s-1\}$, for any two elements $x \neq y \in C_i$, we have $x = \gamma^{lj+i}$ and $y = \gamma^{lj'+i}$ for some $0 \leq j \neq j' \leq s-1$. Since $(r,s) = 1$, we obtain $A_i x^r = A_i \gamma^{lrj+ir} \neq A_i y^r = A_i \gamma^{lrj'+ir}$. Moreover, it is easy to prove that $C_0^r = C_0$ and more generally $C_i^r = C_{ir}$ for any $0 \leq i \leq l-1$. Hence (b) and (c) are equivalent.

Because $A_i \gamma^{ir}$ is a coset representative of $A_i C_{ir}$, it is easy to see that (c), (d), (e), and (f) are equivalent. Finally, since all of $A_0^s, A_1^s\zeta^r, \cdots, A_{l-1}^s\zeta^{(l-1)r}$ are l-th roots of unity, (f) means that $A_0^s, A_1^s\zeta^r, \cdots, A_{l-1}^s\zeta^{(l-1)r}$ are all distinct. By Lemma 2, (f) is equivalent to (g). $\qquad\square$

We note that the equivalence of (a) and (d) was first found in [24] and the equivalence of (a) and (g) was first proved in [4]. In fact, $P(x) = x^r f(x^s)$ is a PP of $\mathbb{F}_q$ if and only if $(r,s) = 1$, $A_i = f(\zeta^i) \neq 0$ for all $0 \leq i \leq l-1$, and any one of the conditions in Theorem 1 holds. From now on, permutation polynomials $P(x) = x^r f(x^s) = f^r_{A_0, A_1, \cdots, A_{l-1}}(x)$ where $A_i = f(\zeta^i)$ for $0 \leq i \leq l-1$ are called *r-th order cyclotomic mapping permutation polynomials of index l*. In the following, we use Theorem 1 (e) to obtain the number of r-th order cyclotomic mapping permutation polynomials of $\mathbb{F}_q$ of index l as in Theorem 2 [20]. The second part is obtained by using Möbius inversion formula of $\sum_{d|l} Q_d = P_l$.

Corollary 1. *Let p be prime, $q = p^m$, and $l \mid q - 1$ for some positive integer l. For each positive integer r such that $(r,s) = 1$, there are $P_l = l!(\frac{q-1}{l})^l$ distinct r-th order cyclotomic mapping permutation polynomials of $\mathbb{F}_q$ of index l. Moreover, the number Q_l of r-th order cyclotomic mapping permutation polynomials of $\mathbb{F}_q$ of least index l is*

$$Q_l = \sum_{d|l} \mu\left(\frac{l}{d}\right) \left(\frac{q-1}{d}\right)^d d!.$$

Therefore there are $l!(\frac{q-1}{l})^l \phi(\frac{q-1}{l})$ distinct permutation polynomials of the form $x^r f(x^s)$ of $\mathbb{F}_q$ in total, which was also obtained in [24] through a study of the group structure of these permutation polynomials.

In the rest of this section, we will see some examples of permutation polynomials of this form. It is well known that $P(x) = x^{r+es}$ is a permutation polynomial of $\mathbb{F}_q$ if and only if $(r + es, q - 1) = 1$, which is equivalent to conditions $(r + es, s) = 1$ and $(r + es, l) = 1$. Obviously, $(r + es, s) = 1$ is the same as $(r, s) = 1$. And the condition $(r + es, l) = 1$ is equivalent to the conditions stated in Theorem 1 for $f(x) = x^e$.

For $l = 3 \mid q - 1$ and an integer $s = \frac{q-1}{3}$, by Theorem 1, $x^r f(x^s)$ is a permutation polynomial of $\mathbb{F}_q$ if and only if $(r,s) = 1$ and $\{A_0^s, A_1^s\theta^r, A_2^s\theta^{2r}\} = \mu_3 = \{1, \theta, \theta^2\}$ where $\theta^3 = 1$ and $A_i = f(\theta^i)$ for $i = 0, 1, 2$. The condition

$\{A_0^s, A_1^s\theta^r, A_2^s\theta^{2r}\} = \{1, \theta, \theta^2\}$ is equivalent to $A_0^s \neq A_1^s\theta^r, A_0^s \neq A_2^s\theta^{2r}, A_1^s\theta^r \neq A_2^s\theta^{2r}$. However, in some cases, we always have $A_0^s = 1$. Next we construct some new classes of permutation polynomials of this type.

Theorem 2. *Let p be prime, $q = p^m$, and $q-1 = 3s$ for some positive integer s. Assume $f(x) \equiv ax^2+bx+c \pmod{x^3-1}$ such that $a^2+b^2+c^2-ab-ac-bc = 1$. Then $P(x) = x^r f(x^s)$ is a permutation polynomial of $\mathbb{F}_q$ if and only if $(r,s) = 1$, $A_0^s = 1$ and $A_1^s\theta^r \neq A_2^s\theta^{2r}$ where $\theta^3 = 1$ and $A_i = f(\theta^i)$ for $i = 0, 1, 2$.*

Proof. If $a^2 + b^2 + c^2 - ab - ac - bc = 1$, then $A_1 A_2 = f(\theta)f(\theta^2) = 1$. If $P(x)$ is a PP, then $\prod_{x \in \mathbb{F}_q^*} P(x) = -1$ implies that $A_0^s = 1$. Hence $P(x) = x^r f(x^s)$ is a permutation polynomial of $\mathbb{F}_q$ if and only if $(r,s) = 1$, $A_0^s = 1$ and $\{A_1^s\theta^r, A_2^s\theta^{2r}\} = \{\theta, \theta^2\}$. Since $A_0^s = 1$, we note that $\{A_1^s\theta^r, A_2^s\theta^{2r}\} = \{\theta, \theta^2\}$ is also equivalent to $A_1^s\theta^r \neq A_2^s\theta^{2r}$. Indeed, $A_0^s = (A_1^s\theta^r)(A_2^s\theta^{2r}) = 1$ implies that either $A_1^s\theta^r = A_2^s\theta^{2r} = 1$ or $\{A_1^s\theta^r, A_2^s\theta^{2r}\} = \{\theta, \theta^2\}$. □

Corollary 2. *Let p be prime, $q = p^m$, and $q - 1 = 3s$ for some positive integer s. Assume that $f(x) \equiv ax^2 + bx + a \pmod{x^3 - 1}$ such that $(a - b)^2 = 1$. Then $x^r f(x^s)$ is a permutation polynomial of $\mathbb{F}_q$ if and only if $(r,s) = 1$, $(2a+b)^s = 1$ and $(r + s, 3) = 1$.*

Proof. Since $f(x) \equiv ax^2 + bx + a \pmod{x^3 - 1}$ and $\theta^3 = 1$, $A_2 = f(\theta^2) = \theta f(\theta) = \theta A_1$. Hence $A_1^s\theta^r \neq A_2^s\theta^{2r}$ reduces to $\theta^{r+s} \neq 1$, which is equivalent to $(r + s, 3) = 1$. The rest follows from Theorem 2 and $A_0 = f(1) = 2a + b$. □

3   Permutation Binomials and Generalized Lucas Sequences

In this section, we explain how permutation binomials and generalized Lucas sequences are closely related as a result of Theorem 1. Again, we let p be prime, $q = p^m$, $q - 1 = ls$ for some positive integers l and s, and ζ be a primitive l-th root of unity. Here we consider permutation polynomials over $\mathbb{F}_q$ of the form $P(x) = x^r(x^{es} + 1)$ with $(e, l) = 1$. That is, $P(x) = x^r f(x^s)$ where $f(x) = x^e + 1$ and $(e, l) = 1$. We note that the case of $f(x) = x^e + b$ with an l-th root of unity b is similar. In this case, p is odd. Otherwise, $P(0) = P(1) = 0$. Moreover, we must have $\zeta^{ei} \neq -1$ for $i = 0, \cdots, l - 1$. Hence l must be odd. Then s must be even. In fact, without loss of generality, we can assume $l \geq 3$ from now on. Moreover, since $l \mid q - 1$ and both p and l are odd, there exists $\eta \in \mathbb{F}_q$ such that $\eta^2 = \zeta$. Hence η is a primitive $2l$-th root of unity in $\mathbb{F}_q$.

We define the sequence $\{a_n\}_{n=0}^{\infty}$ by

$$a_n = \sum_{t=1}^{\frac{l-1}{2}} \left((-1)^{t+1}(\eta^t + \eta^{-t})\right)^n = \sum_{\substack{t=1 \\ t \text{ odd}}}^{l-1} (\eta^t + \eta^{-t})^n.$$

The sequence $\{a_n\}_{n=0}^{\infty}$ is then called *generalized Lucas sequence of order* $\frac{l-1}{2}$ because $\{a_n\}_{n=0}^{\infty} = \{L_n\}_{n=0}^{\infty}$ when $l = 5$, where the Lucas sequence $\{L_n\}_{n=0}^{\infty}$ is an integer sequence satisfying the recurrence relation $L_{n+2} - L_{n+1} - L_n = 0$ and $L_0 = 2$ and $L_1 = 1$.

For any integer $n \geq 1$ and $a \in \mathbb{F}_q$, the Dickson polynomial of the first kind $D_n(x, a) \in \mathbb{F}_q[x]$ of degree n with parameter a is defined by

$$D_n(x, a) = \sum_{i=0}^{\lfloor n/2 \rfloor} \frac{n}{n-i} \binom{n-i}{i} (-a)^i x^{n-2i}.$$

Similarly, the Dickson polynomial of the second kind $E_n(x, a) \in \mathbb{F}_q[x]$ of degree n with parameter a is defined by

$$E_n(x, a) = \sum_{i=0}^{\lfloor n/2 \rfloor} \binom{n-i}{i} (-a)^i x^{n-2i}.$$

For $a \neq 0$, we write $x = y + a/y$ with y an indeterminate. Then the Dickson polynomials can often be rewritten as

$$D_n(x, a) = D_n\left(y + \frac{a}{y}, a\right) = y^n + \frac{a^n}{y^n},$$

and

$$E_n(x, a) = E_n\left(y + \frac{a}{y}, a\right) = \frac{y^{n+1} - a^{n+1}/y^{n+1}}{y - a/y}.$$

In the case $a = 1$, we denote Dickson polynomials of degree n by $D_n(x)$ and $E_n(x)$ respectively. It is well known that Dickson polynomials are closely related to the Chebyshev polynomials by the connections $D_n(2x) = 2T_n(x)$ and $E_n(2x) = U_n(x)$, where $T_n(x)$ and $U_n(x)$ are Chebyshev polynomials of degree n, of the first kind and the second kind respectively. More information on Dickson polynomials can be found in [17].

We consider the Dickson polynomial $E_{l-1}(x)$ of the second kind of degree $l-1$ with parameter $a = 1$. It is well known that $\eta^t + \eta^{-t}$ with $1 \leq t \leq l-1$ are all the roots of $E_{l-1}(x)$ where η is a primitive $2l$-th root of unity. Let

$$E_{l-1}^{odd}(x) = \prod_{\substack{t=1 \\ odd\ t}}^{l-1} (x - (\eta^t + \eta^{-t})).$$

Then the characteristic polynomial of the sequence $\{a_n\}_{n=0}^{\infty}$ is $E_{l-1}^{odd}(x)$. Using the factorization of $U_{l-1}(x)$ over $\mathbb{Z}$ (Theorem 2 in [23]) and the fact $E_{l-1}(2x) = U_{l-1}(x)$, it is obvious to conclude that $E_{l-1}^{odd}(x)$ is also a polynomial with integer coefficients (thus over $\mathbb{F}_p$). It then follows from Waring's formula (Theorem 1.76 in [18]) that $\{a_n\}_{n=0}^{\infty}$ is an integer sequence and thus a sequence over $\mathbb{F}_p$. For more information on the sequence $\{a_n\}_{n=0}^{\infty}$, one can also refer [1], [2], and [3].

Theorem 3. *Let p be odd prime and $q = p^m$. Assume that l, s, r, e are positive integers such that l is odd, $q - 1 = ls$, and $(e, l) = 1$. Then $P(x) = x^r(x^{es} + 1)$ is a permutation polynomial of $\mathbb{F}_q$ if and only if*

(i) $(r, s) = 1$;

(ii) $2^s \equiv 1 \pmod{p}$;

(iii) $2r + es \not\equiv 0 \pmod{l}$;

(iv) $\displaystyle\sum_{k=0}^{cj/2} \frac{cj}{cj-k}\binom{cj-k}{k}(-1)^k a_{cs+cj-2k} = -1$ in $\mathbb{F}_p$ for all $c = 1, \cdots, l-1$, where $\{a_n\}_{n=0}^{\infty}$ is the generalized Lucas sequence of order $\frac{l-1}{2}$ and $2e^{\phi(l)-1}r+s \equiv j \pmod{2l}$.

Proof. Let $P(x)$ be a PP of $\mathbb{F}_q$. It is well-known that $(r, s) = 1$. Moreover, $\displaystyle\prod_{x \in \mathbb{F}_q^*} P(x) = -1$ implies that $\displaystyle\prod_{x \in \mathbb{F}_q^*}(x^{es} + 1) = 1$. Then $\left(\displaystyle\prod_{i=0}^{l-1}(\zeta^i + 1)\right)^s = 1$. Since l is odd, $\displaystyle\prod_{i=0}^{l-1}(\zeta^i + 1) = \prod_{i=0}^{l-1}(1 - (-\zeta^i)) = 1 - (-1) = 2$. Hence $2^s \equiv 1 \pmod{p}$ and (ii) holds.

Assume that $2r + es \equiv 0 \pmod{l}$. Since s is even, $2r + es \equiv 0 \pmod{2l}$. By Theorem 1 (g), we have $\displaystyle\sum_{i=0}^{l-1} \zeta^{cri} A_i^{cs} = 0$ for all $c = 1, \cdots, l-1$, where $A_i = \zeta^{ei}+1$. Since $l \mid q-1$ and l is odd, we can find $\eta \in \mathbb{F}_q$ such that $\eta^2 = \zeta$. Thus we obtain

$$\sum_{i=0}^{l-1} \eta^{(2r+es)ci}(\eta^{ei} + \eta^{-ei})^{cs} = 0, \quad for \; all \; c = 1, \cdots, l-1. \tag{1}$$

It follows from $2r + es \equiv 0 \pmod{2l}$ that

$$\sum_{i=0}^{l-1}(\eta^{ei} + \eta^{-ei})^{cs} = 0, \quad for \; all \; c = 1, \cdots, l-1.$$

Since each $(\eta^{ei} + \eta^{-ei})^s$ is an l-th root of unity, by Lemma 2, $(\eta^{ei} + \eta^{-ei})^s$, $i = 0, \cdots, l-1$, are all distinct. However, since s is even, we have $(\eta^{ei} + \eta^{-ei})^s = (\eta^{e(l-i)} + \eta^{-e(l-i)})^s$, a contradiction. Hence $2r + es \not\equiv 0 \pmod{l}$ and (iii) holds.

Using $(e, l) = 1$, we have $e^{\phi(l)} \equiv 1 \pmod{l}$ where ϕ is the Euler's totient function. Then we can rewrite (1) as

$$2^{cs} + \sum_{t=1}^{(l-1)/2}(\eta^{c(2e^{\phi(l)-1}r+s)t} + \eta^{-c(2e^{\phi(l)-1}r+s)t})(\eta^t + \eta^{-t})^{cs} = 0, \quad for \; all \; c = 1, \cdots, l-1.$$

$$\tag{2}$$

Let $2e^{\phi(l)-1}r+s \equiv j \pmod{2l}$. Then it yields that $\eta^{cjt} + \eta^{-cjt} = D_{cj}(\eta^t + \eta^{-t})$ where $D_{cj}(\eta^t + \eta^{-t})$ is the Dickson polynomial of the first kind of degree cj. That is, $D_{cj}(\eta^t + \eta^{-t}) = \displaystyle\sum_{k=0}^{cj/2} \frac{cj}{cj-k}\binom{cj-k}{k}(-1)^k(\eta^t + \eta^{-t})^{cj-2k}$. Because both s and j are even, we obtain

$$2^{cs} + \sum_{t=1}^{(l-1)/2} (\eta^{c(2e^{\phi(l)-1}r+s)t} + \eta^{-c(2e^{\phi(l)-1}r+s)t})(\eta^t + \eta^{-t})^{cs}$$

$$= 1 + \sum_{t=1}^{(l-1)/2} \sum_{k=0}^{cj/2} \frac{cj}{cj-k} \binom{cj-k}{k} (-1)^k (\eta^t + \eta^{-t})^{cj-2k} (\eta^t + \eta^{-t})^{cs}$$

$$= 1 + \sum_{t=1}^{(l-1)/2} \sum_{k=0}^{cj/2} \frac{cj}{cj-k} \binom{cj-k}{k} (-1)^k (\eta^t + \eta^{-t})^{cj-2k} (\eta^t + \eta^{-t})^{cs}$$

$$= 1 + \sum_{k=0}^{cj/2} \frac{cj}{cj-k} \binom{cj-k}{k} (-1)^k a_{cs+cj-2k}$$

$$= 0.$$

Hence (iv) follows. Conversely, assume that (i)-(iv) are satisfied, then it is straightforward to show that $P(x) = x^r(x^{es} + 1)$ is a permutation polynomial of $\mathbb{F}_q$ by Theorem 1.   □

We can also rewrite the above theorem in the following way.

Corollary 3. *Let* $q = p^m$, p *is an odd prime, and* $q-1 = ls$ *for positive integers* l *and* s. *Assume that* p, l, r, s *satisfies*

$$l \text{ is odd}, (e, l) = 1, (r, s) = 1, 2^s \equiv 1 \pmod{p}, 2r + es \not\equiv 0 \pmod{l}.$$

Then $P(x) = x^r(x^{es} + 1)$ *is a permutation polynomial of* $\mathbb{F}_q$ *if and only if*

$$\sum_{n=0}^{j_c} t_n^{(j_c)} a_{cs+n} = -1 \text{ in } \mathbb{F}_p, \text{ for all } c = 1, \cdots, l-1,$$

where $\{a_n\}_{n=0}^{\infty}$ *is the generalized Lucas sequence of order* $\frac{l-1}{2}$, $2e^{\phi(l)-1}r + s \equiv j \pmod{2l}$, $j_c = cj \bmod 2l$ *and* $t_n^{(j_c)}$ *is the coefficient of* x^n *in the Dickson polynomial of the first kind of degree* j_c.

Furthermore, if $(2r + es, l) = 1$, then we let j' be the inverse of $2e^{\phi(l)-1}r + s$ mod l. Then (2) can be rewritten as

$$2^{c'j's} + \sum_{t=1}^{(l-1)/2} (-1)^{c't}(\eta^{c't}+\eta^{-c't})(\eta^t+\eta^{-t})^{c'j's} = 0, \text{ for all } c' = 1, \cdots, l-1. \quad (3)$$

Using (3) and similar arguments as before, we can improve the previous result by using Dickson polynomials of the first kind of smaller degrees. We note that if c' is even, then the coefficient $t_n^{(c')}$ of x^n in the Dickson polynomial of the first kind of degree c' is always 0 for all odd n. Similarly, if c' is odd, then $t_n^{(c')} = 0$ for all even n.

Corollary 4. *Let* $q = p^m$, p *is odd prime, and* $q - 1 = ls$ *for positive integers* l *and* s. *Assume that* p, l, r, s *satisfies*

$$l \text{ is odd}, (e, l) = 1, (r, s) = 1, 2^s \equiv 1 \pmod{p}, (2r + es, l) = 1.$$

Then $P(x) = x^r(x^{es} + 1)$ *is a permutation polynomial of* $\mathbb{F}_q$ *if and only if*

$$\sum_{n=0}^{c'} t_n^{(c')} a_{c'j's+n} = (-1)^{c'+1} \text{ in } \mathbb{F}_p, \text{ for all } c' = 1, \cdots, l - 1,$$

where $\{a_n\}_{n=0}^{\infty}$ *is the generalized Lucas sequence of order* $\frac{l-1}{2}$, $j'(2e^{\phi(l)-1}r + s) \equiv 1 \pmod{l}$, $t_n^{(c')}$ *is the coefficient of* x^n *in the Dickson polynomial of the first kind of degree* c'.

In particular, when l is a small prime (e.g. $l = 3, 5, 7$), the sequences $\{a_n\}$ that correspond to permutation binomials can be further described explicitly by using the above conditions (see [2], [25]).

References

1. Akbary, A., Alaric, S., Wang, Q.: On some classes of permutation polynomials. Int. J. Number Theory (to appear)
2. Akbary, A., Wang, Q.: On some permutation polynomials. Int. J. Math. Math. Sci. 16, 2631–2640 (2005)
3. Akbary, A., Wang, Q.: A generalized Lucas sequence and permutation binomials. Proc. Amer. Math. Soc. 134(1), 15–22 (2006)
4. Akbary, A., Wang, Q.: On polynomials of the form $x^r f(x^{(q-1)/l})$. Int. J. Math. Math. Sci. (to appear)
5. Bell, J., Wang, Q.: A note on Costas arrays and cyclotomic permutations. Ars Combin. (to appear)
6. Blackburn, S.R., Etzion, T., Paterson, K.G.: Permutation polynomials, de Bruijn sequences, and linear complexity. J. Combin. Theory Ser. A 76(1), 55–82 (1996)
7. Chu, W., Colbourn, C.J., Dukes, P.: Constructions for permutation codes in powerline communications. Des. Codes Cryptogr. 32(1-3), 51–64 (2004)
8. Chu, W., Golomb, S.W.: Circular Tuscan-k arrays from permutation binomials. J. Combin. Theory Ser. A 97(1), 195–202 (2002)
9. Cohen, S.D.: Permutation group theory and permutation polynomials. In: Algebras and combinatorics (Hong Kong, 1997), pp. 133–146. Springer, Singapore (1999)
10. Evans, A.B.: Cyclotomy and orthomorphisms: A survey. Congr. Numer. 101, 97–107 (1994)
11. Golomb, S.W., Moreno, O.: On periodicity properties of Costas arrays and a conjecture on permutation polynomials. IEEE Trans. Inform. Theory 42(6) part 2, 2252–2253 (1996)
12. Levine, J., Brawley, J.V.: Some cryptographic applications of permutation polynomials. Cryptologia 1, 76–92 (1977)
13. Levine, J., Chandler, R.: Some further cryptographic applications of permutation polynomials. Cryptologia 11, 211–218 (1987)

14. Lidl, R., Müller, W.B.: Permutation polynomials in RSA-cryptosystems. In: Adv. in Cryptology, Plenum, New York, pp. 293–301 (1984)
15. Lidl, R., Mullen, G.L.: When does a polynomial over a finite field permute the elements of the field? Amer. Math. Monthly 95, 243–246 (1988)
16. Lidl, R., Mullen, G.L.: When does a polynomial over a finite field permute the elements of the field? II. Amer. Math. Monthly 100, 71–74 (1993)
17. Lidl, R., Mullen, G.L., Turnwald, G.: Dickson polynomials. In: Pitman Monographs and Surveys in Pure and Applied Math., Longman, London/Harlow/Essex (1993)
18. Lidl, R., Niederreiter, H.: Finite Fields. In: Encyclopedia of Mathematics and Its Applications, Cambridge University Press, Cambridge (1997)
19. Mullen, G.L.: Permutation polynomials over finite fields. In: Finite Fields, Coding Theory, and Advances in Communications and Computing, pp. 131–151. Marcel Dekker, New York (1993)
20. Niederreiter, H., Winterhof, A.: Cyclotomic $\mathcal{R}$-orthomorphisms of finite fields. Discrete Math. 295, 161–171 (2005)
21. Rivest, R.L.: Permutation polynomials modulo 2^w. Finite fields Appl. 7, 287–292 (2001)
22. Rivest, R.L., Robshaw, M.J.B., Sidney, R., Yin, Y.L.: The RC6 block cipher, Published electoronically at http://theory.lcs.mit.edu/rivest/rc6.pdf
23. Rayes, M.O., Trevisan, V., Wang, P.: Factorization of Chebyshev polynomials, http://icm.mcs.kent.edu/reports/index1998.html
24. Wan, D., Lidl, R.: Permutation polynomials of the form $x^r f(x^{(q-1)/d})$ and their group structure. Monatsh. Math. 112, 149–163 (1991)
25. Wang, L.: On permutation polynomials. Finite Fields Appl. 8, 311–322 (2002)
26. Sun, J., Takeshita, O.Y.: Interleavers for Turbo codes using permutation polynomials over integer rings. IEEE Trans. Inform. Theory 51(1), 101–119 (2005)

Single-Track Gray Codes and Sequences

Tuvi Etzion

Technion — Israel Institute of Technology
Department of Computer Science
Technion City, Haifa 32000, Israel
etzion@cs.technion.ac.il

Abstract. Single-track Gray codes are cyclic Gray codes with code-words of length n, such that all the n tracks which correspond to the n distinct coordinates of the codewords are cyclic shifts of the first track. These codes have advantages over conventional Gray codes in certain quantization and coding applications. We survey the main results on this problem, their connections to sequence design, and discuss some of the interesting open problems.

1 Introduction

Gray codes were found by Gray [3] and introduced by Gilbert [2] as a listing of all binary n-tuples in a list such that any two consecutive n-tuples in the list differ in exactly one position. Generalization of Gray codes were given during the years. They include listing subsets of the binary n-tuples in a Gray code manner, in such a way that the list has some prespecified properties. These properties were usually forced by a specific application for the Gray code. An excellent survey on Gray code is given in [6].

A *single-track* Gray code is a list of P distinct binary words of length n, such that each two consecutive words, including the last and the first, differ in exactly one position and when looking at the list as an $P \times n$ array, each column of the array is a cyclic shift of the first column. These codes were defined by Hiltgen, Paterson, and Brandenstini [4] who also gave their main application. A length n, period P Gray code can be used to record the absolute angular positions of a rotating wheel by encoding the codewords on n concentrically arranged tracks. Then n reading heads, mounted in parallel across the tracks suffice to recover the codewords. When the heads are nearly aligned with the division between two codewords, any components which change between those words will be in doubt and a spurious position value will result. Such quantization errors are minimized by using a Gray code encoding, for then exactly one component can be in doubt and the two codewords that could possibly result identify the positions bordering the division, resulting in a small angular error. When high resolution is required, the need for a large number of concentric tracks results in encoders with large physical dimensions. This poses a problem in the design of small-scale or high-speed devices. Single-track Gray codes were proposed in [4] as a way of overcoming these problems.

S.W. Golomb et al. (Eds.): SSC 2007, LNCS 4893, pp. 129–133, 2007.

The aim of this note is to survey the known results on constructions for single-track Gray codes, the properties of these codes, and nonexistence results, as well as the main open problems in this topic.

2   Constructions, Properties, and Nonexistence Results

Let $W = [w_0, w_1, \ldots, w_{n-1}]$ be a length n word. The *cyclic shift operator*, $\mathbf{E}$, is defined by $\mathbf{E}W = [w_1, w_2, \ldots, w_{n-1}, w_0]$ and the *complementary cyclic shift operator* $\bar{\mathbf{E}}$ is defined by $\bar{\mathbf{E}}W = [w_1, w_2, \ldots, w_{n-1}, \bar{w}_0]$, where $\bar{b}$ is the binary complement of b.

Let W be a length n word. The *cyclic order* of W is defined as

$$o(W) = \min\{i \; : \; E^i W = W, \; i \geq 1\}$$

The *complementary cyclic order* of W is defined as

$$\bar{o}(W) = \min\{i \; : \; \bar{E}^i W = W, \; i \geq 1\}$$

If $o(W) = n$ then W has *full-order* and if $\bar{o}(W) = 2n$ then $W\bar{W}$ is a *full-order self-dual* word.

Let $\mathcal{C}$ be an ordered list of P length n

$$W_0, W_1, \ldots, W_{P-1}.$$

For each $0 \leq i < P$ we denote the components of W_i as

$$W_i = [w_i^0, w_i^1, \ldots, w_i^{n-1}].$$

The jth track of $\mathcal{C}$, for $0 \leq j < n$, is defined as

$$t_j(\mathcal{C}) = [w_0^j, w_1^j, \ldots, w_{P-1}^j]$$

$\mathcal{C}$ has the *single-track property* if there exist integers

$$k_0, k_1, \ldots, k_{n-1}$$

called the *head positions*, where $k_0 = 0$, such that $t_i(\mathcal{C}) = \mathbf{E}^{k_i} t_0(\mathcal{C})$ for each $0 \leq i < n$. For each $0 \leq i < n$, k_i is called the *position of the ith head*.

Let $\mathcal{C}$ be an ordered list of P length n

$$W_0, W_1, \ldots, W_{P-1}.$$

$\mathcal{C}$ is a length n, period P single-track Gray code if $\mathcal{C}$ is a cyclic Gray code and $\mathcal{C}$ has the single-track property.

We can describe a length n period P Gray code as a sequence of P integers taken from the set $\{1, 2, \ldots, n\}$, where the consecutive integers point on the number of the coordinate in which the two corresponding rows. The following two properties are easy to verify:

- The number of times each integer appears in the sequence is even.
- Any two integers appear the same number of times in the sequence.

These properties lead to the following theorem [4].

Theorem 1. *For a length n period P single-track Gray code we have $2n$ divides P and $2n \leq P \leq 2^n$.*

Necklaces ordering

Let $S_0, S_1 \ldots, S_{r-1}$ be r length n binary pairwise nonequivalent full-order words, such that for each $0 \leq i < r - 1$, S_i, and S_{i+1} differ in exactly one coordinate, and there also exists an integer ℓ, $gcd(\ell, n) = 1$, such that S_{r-1} and $E^\ell S_0$ differ in exactly one coordinate, then the following words form a length n, period nr single-track Gray code

$$
\begin{array}{cccc}
S_0, & S_1, & \cdots & S_{r-1} \\
E^\ell S_0, & E^\ell S_1, & \cdots & E^\ell S_{r-1} \\
E^{2\ell} S_0, & E^{2\ell} S_1, & \cdots & E^{2\ell} S_{r-1} \\
\vdots & \vdots & \vdots & \vdots \\
E^{(n-1)\ell} S_0, & E^{(n-1)\ell} S_1, & \cdots & E^{(n-1)\ell} S_{r-1}
\end{array}
$$

Self-dual necklaces ordering

Let $S_0, S_1 \ldots, S_{r-1}$ be r length $2n$ self-dual pairwise nonequivalent full-order words.

For each i, $1 \leq i \leq r - 1$, let $S_i = [s_i^0, s_i^1, \ldots, s_i^{2n-1}]$ and let

$$
F^j S_i = [s_i^j, s_i^{j+1}, \ldots, s_i^{j+2n-1}]
$$

where superscripts are taken modulo $2n$.

If for each $0 \leq i < r - 1$, S_i, and S_{i+1} differ in exactly two coordinates, and there also exists an integer ℓ, $gcd(\ell, 2n) = 1$, such that S_{r-1} and $E^\ell S_0$ differ in exactly one coordinate, then the following words form a length n, period $2nr$ single-track Gray code

$$
\begin{array}{cccc}
F^0 S_0, & F^0 S_1, & \cdots & F^0 S_{r-1} \\
F^\ell S_0, & F^\ell S_1, & \cdots & F^\ell S_{r-1} \\
F^{2\ell} S_0, & F^{2\ell} S_1, & \cdots & F^{2\ell} S_{r-1} \\
\vdots & \vdots & \vdots & \vdots \\
F^{(2n-1)\ell} S_0, & F^{(2n-1)\ell} S_1, & \cdots & F^{(2n-1)\ell} S_{r-1}
\end{array}
$$

Let $\mathcal{C}$ be a length n, period P single-track Gray code, and let the head positions be $k_0, k_1, \ldots, k_{n-1}$. $\mathcal{C}$ has *k-spaced heads* if

$$k_{i+1} \equiv k_i + k \pmod{P}$$

for each $0 \leq i \leq n - 2$.

Let $\mathcal{C}$ be a set of words. The *cyclic order* and *complementary cyclic order* of $\mathcal{C}$ are defined as

$$o(\mathcal{C}) = \min\{o(W) \; : \; W \in \mathcal{C}\}$$

$$\bar{o}(\mathcal{C}) = \min\{\bar{o}(W) \; : \; W \in \mathcal{C}\}$$

The following three results on single-track Gray codes were given in [5].

Theorem 2. *Let $\mathcal{C}$ be a length n period P single-track Gray code with k-spaced heads.*

- *If k is even then*
  - $gcd(k, P) = \frac{P}{o(\mathcal{C})}$.
  - $o(W) = o(\mathcal{C}) = n$ *for each $W \in \mathcal{C}$.*
  - *There exists an ordering of $\frac{P}{o(\mathcal{C})}$ length n necklace representatives of cyclic order n, which satisfies the requirements of necklaces ordering.*

- *If k is odd then*
  - $gcd(k, P) = \frac{P}{\bar{o}(\mathcal{C})}$.
  - $\bar{o}(W) = \bar{o}(\mathcal{C}) = 2n$ *for each $W \in \mathcal{C}$.*
  - *There exists an ordering of $\frac{P}{\bar{o}(\mathcal{C})}$ length $2n$ self-dual necklace representatives of cyclic order $2n$, which satisfies the requirements of self-dual necklaces ordering.*

Lemma 1. *If $\mathcal{C}$ is a length n period P single-track Gray code with k-spaced heads, k odd, then each track of the code is self-dual.*

Theorem 3. *There is no ordering of all the 2^n words of length $n = 2^m$, $m \geq 2$, in a list which satisfies the following three requirements:*

- *Each two adjacent words have different parity.*
- *The list has the single-track property,*
- *each word appears exactly once.*

Etzion and Paterson [1] gave several constructions of single-track Gray codes obtained by necklaces ordering. One construction produces a special arrangement of $2^{n-1}r$ nonequivalent full-order words of length $2n$, which satisfies certain conditions (which are usually not difficult to satisfy), from a special arrangement of r full-order words of length n which satisfies the same conditions. But, their most impressive construction is optimal single-track Gray codes of length 2^m, $m \geq 3$, which attain the bound implied by the nonexistence result of Theorem 3.

Theorem 4. *There exists an ordering of the self-dual necklaces of length $2n = 2^{m+1}$, $m \geq 3$, from which we obtain a length $n = 2^m$ period $2^n - 2n$ single-track Gray code.*

Schwartz and Etzion [5] gave another recursive construction based on the existing of two single-track Gray of length n and period $2^n - c_n$ and length k and period $2^k - c_k$, obtained by necklaces ordering, with certain properties (which are usually not difficult to satisfy). The construction which is again obtained by necklaces ordering has length nk and period $2^{nk} - c_{nk}$ where

$$c_{nk} = 2^{nk}(c_k 2^{-k} + c_n 2^{-n} - c_k c_n 2^{-(n+k)}).$$

If we further assume that we have sequences of single-track Gray codes such that

$$\lim_{n \to \infty} \frac{c_n}{2^n} = 0 \qquad \lim_{k \to \infty} \frac{c_k}{2^k} = 0$$

then we have

$$\lim_{n,k \to \infty} \frac{c_{nk}}{2^{nk}} = 0.$$

3   Open Problems

The constructions for large period single-track Gray code require seed-codes with large period. Constructions of such seed-codes is the major open problem in this topic. If $n = 2^m$ then a construction of code with period $2^n - 2n$ is known, but there is no code with a larger period. The other values of n which is of special interest are those when n is an odd prime number. In this case there are exactly $\frac{2^n - 2}{n}$ pairwise nonequivalent full-order words of length n. Can they be ordered to form a period $2^n - 2$ single-track Gray code? We conjecture that the answer is yes. This conjecture was verified to be true if $n \leq 13$ [1] and we verified that it is also true if n equals either 17 or 19. A general construction of this sort will be a major breakthrough.

References

1. Etzion, T., Paterson, K.G.: Near optimal single-track Gray codes. IEEE Trans. on Inform. Theory 42, 779–789 (1996)
2. Gilbert, E.N.: Gray codes and paths on the n-cube. Bell Syst. Tech. Journal 37, 815–826 (1958)
3. Gray, F.: Pulse codes. US Patent application 237(7), 94–111 (1953)
4. Hiltgen, A.P., Paterson, K.G., Brandestini, M.: Single-track Gray codes. IEEE Trans. on Inform. Theory 42, 1555–1561 (1996)
5. Schwartz, M., Etzion, T.: The structure of single-track Gray codes. IEEE Trans. on Inform. Theory 45, 2383–2396 (1999)
6. Savage, C.: A survey of combinatorial Gray codes. SIAM Reviews 39, 605–629 (1997)

The Asymptotic Behavior of π-Adic Complexity with $\pi^2 = -2$

Andrew Klapper[*]

Dept. of Computer Science
779A Anderson Hall
University of Kentucky
Lexington, KY, 40506-0046
www.cs.uky.edu/~klapper

Abstract. We study the asymptotic behavior of stream cipher security measures associated with algebraic feedback shift registers and feedback based on the ring $\mathbb{Z}[\sqrt{-2}]$. For non-periodic sequences we consider normalized $\sqrt{2}$-adic complexity and study the set of accumulation points for a fixed sequence. The the set of accumulation points is a closed subinterval of the real closed interval $[0,1]$. We see that this interval is of the form $[B, 1-B]$ "most" of the time, and that all such intervals occur for some sequence.

Keywords: Sequences, N-adic complexity, Stream ciphers, shift registers.

1 Introduction

The purpose of this paper is to study the asymptotic behavior of security or randomness measures for infinite sequences. The kinds of measures we are interested in arise in the following manner. There is a class $\mathcal{G}$ of finite state devices that generate infinite sequences over some alphabet Σ, such that every eventually periodic sequence is generated by at least one element of $\mathcal{G}$. We also assume there is a notion of the size of a generator in $\mathcal{G}$, a positive real number. In general this measure should be close to the number, n, of elements of Σ needed to represent a state of the generator. Typically "close" means differing from n by at most $O(\log(n))$. Examples of such classes of generators include the *linear feedback shift registers* (LFSRs), where the size of an LFSR is the number of cells, and *feedback with carry shift registers* (FCSRs) [7], where the size of an FCSR is the log of the connection number (the log with base equal to the size of the output alphabet).

[*] Parts of this work were carried out while the author was visiting the Fields Institute at the University of Toronto; The Institute for Advanced Study; and Princeton Universiy. This material is based upon work supported by the National Science Foundation under Grant No. CCF-0514660. Any opinions, findings, and conclusions or recommendations expressed in this material are those of the authors and do not necessarily reflect the views of the National Science Foundation.

S.W. Golomb et al. (Eds.): SSC 2007, LNCS 4893, pp. 134–146, 2007.

We denote by $\lambda^{\mathcal{G}}(S)$ the minimum size of a generator in $\mathcal{G}$ that outputs the eventually periodic sequence S. In many cases there is an algorithm that efficiently finds the minimum size generator of S given a prefix of S whose length is a small (typically linear) function of $\lambda^{\mathcal{G}}(S)$. We call this a $\mathcal{G}$-*synthesis algorithm*. Examples include the Berlekamp-Massey algorithm for LFSRs [10] and the 2-adic rational approximation algorithm for FCSRs [7]. When a $\mathcal{G}$-synthesis algorithm exists, the quantity $\lambda^{\mathcal{G}}(S)$ is a measure of the security of S.

Many sequences cannot be generated by a generator in $\mathcal{G}$. For LFSRs and AFSRs, these are exactly the sequences that are not ultimately periodic (but this is not so in general). For such sequences S the measure $\lambda^{\mathcal{G}}(S)$ is undefined. However, we can apply the measure to the various prefixes of S and try to understand the asymptotic behavior. For $n > 0$, let $\lambda_n^{\mathcal{G}}(S)$ denote the minimum size of a generator from $\mathcal{G}$ that outputs the first n symbols of S as its first n outputs. The sequence of numbers $(\lambda_n^{\mathcal{G}}(S) : n = 1, 2, \cdots)$ is called the $\mathcal{G}$-*complexity profile of S*. For a sequence that cannot be generated by a generator in $\mathcal{G}$, the limit of the $\lambda_n^{\mathcal{G}}(S)$ is infinite, so we normalize the measure by letting $\delta_n^{\mathcal{G}}(S) = \lambda_n^{\mathcal{G}}(S)/n$. For the typical measures we are interested in we have

$$\lambda_n^{\mathcal{G}}(S) \leq n + O(\log(n)),$$

so that

$$0 \leq \delta_n^{G}(S) \leq 1 + o(n).$$

In particular, the limsup of the $\delta_n^{G}(S)$ is at most 1. In general the $\delta_n^{\mathcal{G}}(S)$ do not have a single limit, but rather have a set $T(S)$ of accumulation points. Our goal is to determine what sets of accumulation points are possible.

It is immediate for such a measure $\lambda_n^{\mathcal{G}}(S)$ that

$$\lambda_{n+1}^{\mathcal{G}}(S) \geq \lambda_n^{\mathcal{G}}(S)$$

for all $n \geq 1$, so that

$$\delta_{n+1}^{\mathcal{G}}(S) \geq \frac{n}{n+1}\delta_n^{\mathcal{G}}(S) \geq \delta_n^{\mathcal{G}}(S) - \frac{1}{n+1}. \tag{1}$$

This allows us to show that the set of accumulation points is a closed interval [5].

Theorem 1. *Let* $\{\lambda_n : n = 1, 2, \infty\}$ *be a sequence of integers,* $1 \leq \lambda_n \leq n$, *satisfying* $\lambda_n \leq \lambda_{n+1}$ *for all* $n = 1, 2, \cdots$. *Let* $\delta_n = \lambda_n/n \in [0, 1]$. *Then the set* T *of accumulation points of the* δ_n *is a closed real interval* $[B, C] = \{a \in \mathbb{R} : B \leq a \leq C\}$.

Dai, Jiang, Imamura, and Gong studied this problem in the case when $\mathcal{G}$ is the set of LFSRs over $\mathbb{F}_2$, the finite field with two elements, and $\lambda^{\mathcal{G}}$ is linear complexity [3]. They showed that in this case $T(S)$ is an interval of the form $[B, 1 - B]$, with $0 \leq B \leq 1/2$. They also showed that for every such B there are sequences S with $T(S) = [B, 1 - B]$. Dai, Imamura, and Yang, and Feng and Dai have also studied this problem for vector valued non-periodic sequences [2,4]. In this setting, however, there are much more limited results. They showed that there is a number associated with the generalized continued fraction expansion of a multisequence that is a lower bound for the maximum accumulation point and an upper bound for the minimum accumulation point. The author of the current paper studied this problem for FCSRs based on N-adic numbers, where N is a positive integer [5]. He showed that if N is a power of 2 or 3, then $T(S)$ is an interval of the form $[B, 1 - B]$, with $0 \leq B \leq 1/2$. However, for more general N it was only shown that $T(S)$ is an interval of this form if the least accumulation point B satisfies $B \leq \log_N(2)$. It is unknown what can happen if $B > \log_N(2)$.

The primary goal of this paper is to do a similar analysis for π-adic complexity, where $\pi^2 = -p$ and p is a positive square-free integer greater than 1. Let $R = \mathbb{Z}[\pi] = \mathbb{Z} + \mathbb{Z}\pi$.

Consider the alphabet $\Sigma = \{0, 1, \cdots, p - 1\}$. Recall that π-adic complexity is the security measure for p-ary sequences associated with the set $\mathcal{G}$ of π-ary algebraic feedback shift registers (AFSRs) [7,8,9]. Such a register (with a particular initial state) can be identified with an element a/b with $a, b \in R = \mathbb{Z}[\pi]$ in much the same way that a linear feedback shift register can be identified with a rational function $a(x)/b(x)$. The output sequence is then the π-adic expansion of a/b,

$$\frac{a}{b} = \sum_{i=0}^{\infty} s_i \pi^i. \tag{2}$$

We denote the set

$$\left\{ \sum_{i=0}^{\infty} s_i \pi^i : s_i \in \{0, 1, \cdots, N - 1\} \right\}$$

of π-adic numbers by R_π. It is well-known that R_π is an algebraic ring. We have the usual algebraic norm function N from R to $\mathbb{Z}$: if $u, v \in \mathbb{Z}$ then $N(u + v\pi) = u^2 + pv^2$ [1]. For a pair of elements $x, y \in R$, let $\Phi(x, y) = \max(N(x), N(y))$ and $\lambda(x, y) = \log_p(\Phi(x, y))$.

Let D be an AFSR that outputs a sequence S from initial state σ. We write $D(\sigma) = S$. Let $\mu(D, \sigma)$ denote the number of p-ary cells required to represent all states in the infinite execution of D with initial state σ.

Theorem 2. *Let S be a p-ary sequence generated by an AFSR D over R and π with connection element b. Then $\sum_{i=0}^{\infty} s_i \pi^i = a/b$, for some $a \in R$. We have $|\mu(D, \sigma) - \lambda(a, b)| \leq \log_p(\lambda(a, b))$.*

We take $\lambda(a, b)$ as a measure of the size of the generator of S. It is a security measure in the sense that p-ary sequences generated by AFSRs with small $\lambda(u, q)$ are insecure. We further let $\Phi_n(S)$ denote the least $\Phi(a, b)$ such that

$$\frac{a}{b} \equiv \sum_{i=0}^{\infty} s_i \pi^i \pmod{\pi^n}$$

and let $\lambda_n(S) = \log_p(\Phi_n(S))$. Let $\delta_n(S) = \lambda_n(S)/n$. Also, we let $\Phi(S)$ denote the least $\Phi(a, b)$ so that equation (2) holds, if such a pair a, b exists. We let

$\Phi(S) = \infty$ otherwise. Let $\lambda(S) = \log_p(\Phi(S))$, the π-*adic complexity of* S. The sequence of numbers $(\lambda_1(S), \lambda_2(S), \cdots)$ is called the π-*adic complexity profile of* S. We can take

$$a = \sum_{i=0}^{n-1} s_i \pi^i, \quad b = 1$$

so that $a/b \equiv \sum_{i=0}^{\infty} s_i \pi^i \pmod{\pi^n}$. Let $a = a_0 + a_i \pi$ with $a_i \in \mathbb{Z}$. Then

$$a_0 \leq \begin{cases} p^{n/2} - 1 & \text{if } n \text{ is even} \\ p^{(n+1)/2} - 1 & \text{if } n \text{ is odd,} \end{cases} \quad \text{and} \quad a_1 \leq \begin{cases} p^{n/2} - 1 & \text{if } n \text{ is even} \\ p^{(n-1)/2} - 1 & \text{if } n \text{ is odd.} \end{cases}$$

Thus

$$\Phi_n(S) \leq N(a) = \begin{cases} (p+1)(p^{n/2} - 1)^2 & \text{if } n \text{ is even} \\ (p^{(n+1)/2} - 1)^2 + p(p^{(n-1)/2} - 1)^2 & \text{if } n \text{ is odd.} \end{cases} \leq (p+1)p^n.$$

Therefore $\lambda_n(S) < \log_p((p+1)p^n) = n + \log_p(p+1)$ and $\limsup(\delta_n(S)) \leq 1$. There is an effective register synthesis algorithm for these AFSRs [9], so π-adic complexity is an important security measure for p-ary sequences.

2   Basic Lemmas

Let p be a positive square free integer and let $\pi^2 = -p$. Our goal is to find the structure of the set $T_\pi(S)$ of accumulation points of the normalized π-adic complexity profile of an ultimately nonperiodic sequence S. In this section we develop technical tools do this.

If S is ultimately periodic, then $\Phi_n(S)$ is constant for n sufficiently large, so the limit of the $\delta_n(S)$ exists and is zero. Therefore from here on in this section we assume that S is not ultimately periodic.

Lemma 1. *For any* $a, b \in R$ *we have*

1. $N(a) \geq 0$ *and* $N(a) = 0$ *if and only if* $a = 0$.
2. $N(a) = 1$ *if and only if* a *is a unit in* R *if and only if* $a \in \{1, -1\}$.
3. $N(\pi) = p$.
4. $N(ab) = N(a)N(b)$.
5. $N(a + b) \leq N(a) + N(b) + 2(N(a)N(b))^{1/2} = (N(a)^{1/2} + N(b)^{1/2})^2 \leq 4\max(N(a), N(b))$.

Proof: All but the last statement are standard. Let $a = a_0 + a_1 \pi$ and $b = b_0 + b_1 \pi$, $a_0, a_1, b_0, b_1 \in \mathbb{Z}$. Then $N(a + b) = (a_0 + b_0)^2 + p(a_1 + b_1)^2 = N(a) + N(b) + 2(a_0 b_0 + p a_1 b_1)$. We have $0 \leq (a_1 b_0 - a_0 b_1)^2 = a_1^2 b_0^2 + a_0^2 b_1^2 - 2 a_0 a_1 b_0 b_1$. Thus $2 a_0 a_1 b_0 b_1 \leq a_1^2 b_0^2 + a_0^2 b_1^2$. Therefore

$$\begin{aligned} (a_0 b_0 + p a_1 b_1)^2 &= a_0^2 b_0^2 + 2 p a_0 a_1 b_0 b_1 + p^2 a_1^2 b_1^2 \\ &\leq a_0^2 b_0^2 + p(a_1^2 b_0^2 + a_0^2 b_1^2) + p^2 a_1^2 b_1^2 \\ &= N(a)N(b). \end{aligned}$$

Hence $a_0 b_0 + p a_1 b_1 \leq (N(a)N(b))^{1/2}$ and the lemma follows.   □

Lemma 2. *If $p = 2$, then for every $a, b \in R$ with $b \neq 0$, there exist $q, r \in R$ so that $a = qb + r$ and $N(r) < (3/4)N(b)$.*

In particular R is a Euclidean domain and hence a GCD domain. Thus it makes sense to speak of the greatest common divisor of a pair of elements. Lemma 2 is false for all $p > 2$.

We need a lemma to bound $\delta_{n+1}(S)$ in terms of $\delta_n(S)$.

Lemma 3. *Suppose that $\Phi_{n+1}(S) > \Phi_n(S)$.*

1. We have

$$\frac{p^n}{4\Phi_n(S)} \leq \Phi_{n+1}(S)$$

   Therefore

$$\frac{n - \log_p(4)}{n+1} - \frac{n}{n+1}\delta_n(S) \leq \delta_{n+1}(S).$$

2. If $p = 2$, then

$$\Phi_{n+1}(S) \leq \frac{3\Phi_n(S)}{2} + \frac{2^n}{\Phi_n(S)} + \sqrt{6} \cdot 2^{n/2}.$$

3. If $p = 2$, then for all $\epsilon > 0$, if n is sufficiently large and $\delta_n(S) > \max(1/2 + \epsilon, \log_2(3(1 + 2\epsilon)/2))$, then $\delta_{n+1}(S) < \delta_n(S)$.

Proof: Let

$$\frac{a}{b} \equiv \sum_{i=0}^{\infty} s_i \pi^i \pmod{\pi^n} \tag{3}$$

with b a unit modulo π and $\Phi(a, b) = \Phi_n(S)$. By the assumption that $\Phi_{n+1}(S) > \Phi_n(S)$, equation (3) does not hold modulo π^{n+1}. Thus for some $v \in R$ with v not divisible by π we have

$$\frac{a}{b} \equiv v\pi^n + \sum_{i=0}^{\infty} s_i \pi^i \pmod{\pi^{n+1}}.$$

Suppose also that

$$\frac{c}{d} \equiv \sum_{i=0}^{\infty} s_i \pi^i \pmod{\pi^{n+1}}$$

with d a unit modulo π and $\Phi(c, d) = \Phi_{n+1}(S)$. Then

$$\frac{a}{b} \equiv \frac{c}{d} + v\pi^n \pmod{\pi^{n+1}}.$$

It follows that $ad - bc \equiv bdv\pi^n \pmod{\pi^{n+1}}$. Since v is nonzero, we have

$$p^n = N(\pi^n) \leq N(ad - bc) \leq 4\Phi_n(S)\Phi_{n+1}(S).$$

This implies the lower bound in the first assertion. The lower bound on $\delta_{n+1}(S)$ follows by taking logarithms and dividing by $n + 1$.

To obtain an upper bound on $\Phi_{n+1}(S)$ with $p = 2$, we construct a "pretty good" rational approximation modulo π^{n+1}. Then $\Phi_{n+1}(S)$ is upper bounded by the value of Φ on this approximation. Note that a and b have no nontrivial common divisors: if they did, then we could factor out a common factor and reduce $\Phi(a, b)$. Also, $v = 1$ since $p = 2$.

First assume that $0 < N(b) \le N(a)$. Since $a\pi$ and b have no common divisors and R is a Euclidean domain, there exist $e, f \in R$ so that

$$a\pi e - bf = \pi^n - a. \tag{4}$$

Let $g = 1 + \pi e$, so that $g \equiv 1 \pmod{\pi}$ and

$$ag - bf = \pi^n. \tag{5}$$

It then follows that $ag - bf = \pi^n \equiv bvg\pi^n \pmod{\pi^{n+1}}$, since $b \equiv 1 \pmod{\pi}$. Thus

$$\frac{a}{b} - \frac{f}{g} \equiv \pi^n \pmod{\pi^{n+1}},$$

so

$$\frac{f}{g} \equiv \sum_{i=0}^{\infty} s_i \pi^i \pmod{\pi^{n+1}},$$

and g is a unit modulo π. Therefore $\Phi(f, g)$ is an upper bound for $\Phi_{n+1}(S)$. In fact there are many choices for (f, g) satisfying equation (5). By the relative primality of $a\pi$ and b, the solutions to equation (5) with $g \equiv 1 \pmod{\pi}$ are exactly the pairs $(f, g) = (f_0, g_0) + (ra\pi, rb\pi)$ where (f_0, g_0) is a fixed solution and $r \in R$. In particular, we can take $N(f) < (3/4)N(a\pi) = (3/2)\Phi_n(S)$. We then have $ag = bf + \pi^n$ so that

$$N(g) \le \frac{N(bf) + N(\pi^n) + 2N(bf)^{1/2}N(\pi^n)^{1/2}}{N(a)}$$

$$\le \frac{3N(b)}{2} + \frac{2^n}{N(a)} + \sqrt{6} \cdot 2^{n/2}$$

$$\le \frac{3\Phi_n(S)}{2} + \frac{2^n}{\Phi_n(S)} + \sqrt{6} \cdot 2^{n/2}.$$

Therefore

$$\Phi_{n+1}(S) \le \frac{3\Phi_n(S)}{2} + \frac{2^n}{\Phi_n(S)} + \sqrt{6} \cdot 2^{n/2}. \tag{6}$$

This proves the second assertion when $N(b) \le N(a)$.

Now let $0 < N(a) < N(b)$. As in the previous case there are integers $g = 1 + \pi e$ and f with $ag - bf = \pi^n$. By adding a multiple of $(b\pi, a\pi)$ to the pair (g, f), we may assume that $N(g) < 3N(b\pi)/4 = 3\Phi_n(S)/2$. It follows that

$$N(f) \le \frac{3\Phi_n(S)}{2} + \frac{2^n}{\Phi_n(S)} + \sqrt{6} \cdot 2^{n/2}.$$

Finally we prove the third assertion. Let $\epsilon > 0$ and suppose that $\delta_n(S) > \max(1/2 + \epsilon, \log_2(3(1 + 2\epsilon)/2))$. Take n large enough that

$$n > \frac{\log_2(8/(3\epsilon^2))}{2\epsilon}. \tag{7}$$

From $\delta_n(S) > (1/2) + \epsilon$ and equation (7) it then follows that

$$\frac{2^n}{\Phi_n(S)} < \epsilon \frac{3\Phi_n(S)}{2}$$

and

$$\sqrt{6} \cdot 2^{n/2} < \epsilon \frac{3\Phi_n(S)}{2}.$$

Also, from $\delta_n(S) > \log_2(3(1 + 2\epsilon)/2)$ it follows that

$$(1 + 2\epsilon)\frac{3\Phi_n(S)}{2} < \Phi_n(S)^{(n+1)/n}.$$

It then follows that

$$\Phi_{n+1}(S) \leq \frac{3\Phi_n(S)}{2} + \frac{2^n}{\Phi_n(S)} + \sqrt{6} \cdot 2^{n/2}$$
$$\leq (1 + 2\epsilon)\frac{3\Phi_n(S)}{2}$$
$$< \Phi_n(S)^{(n+1)/n}.$$

Taking logarithms and dividing by $n + 1$ then gives $\delta_{n+1}(S) < \delta_n(S)$ as desired.
□

This will suffice to characterize sets $[B, C]$ of accumulation points of normalized π-adic complexities of sequences when $p = 2$ and

$$1 - B \geq \lim_{\epsilon \to 0} \max\left(\frac{1}{2} + \epsilon, \log_2(3(1 + 2\epsilon)/2)\right) = \max\left(\frac{1}{2}, \log_2(3/2)\right).$$

This is equivalent to having $B \leq \min(1/2, \log_2(4/3)) = \log_2(4/3) \sim 0.415$.

3   Sets of Accumulation Points

Let S be an ultimately non-periodic π-ary sequence. In this section we show that in many cases the set of accumulation points $T_\pi(S)$ satisfies $T_\pi(S) = [B, 1 - B]$ for some B. Let $T_\pi(S) = [B, C]$. Let $m_1, m_2, \cdots$ be a sequence of indices such that $B = \lim_{n \to \infty} \delta_{m_n}(S)$. If $\lambda_{n+1}(S) = \lambda_n(S)$, then $\delta_{n+1}(S) < \delta_n(S)$. If we replace m_n by the next index j so that $\lambda_j(S) < \lambda_{j+1}(S)$, then the resulting sequence will have a limit $D \leq B$. Since B is the minimal accumulation point of the $\delta_i(S)$, $D = B$. Therefore we may assume that $\lambda_{m_n}(S) < \lambda_{m_n+1}(S)$.

Lemma 4. *Let $B < 1/2$. Then*

$$\lim_{n\to\infty} \delta_{m_n+1} \geq 1 - B.$$

If $p = 2$, then

$$\lim_{n\to\infty} \delta_{m_n+1} = 1 - B.$$

Proof: Let $0 < \epsilon < 1/2 - B$. Take n large enough that $m_n \geq 6/\epsilon$ and $|B - \delta_{m_n}(S)| < \min(\epsilon/2, (1 - 2B)/4)$. Then $\delta_{m_n}(S) < 1/2$ and by Lemma 3.1

$$1 - B - \delta_{m_n+1}(S) \leq 1 - B - \left(\frac{m_n - 2}{m_n + 1} - \frac{m_n}{m_n + 1} \delta_{m_n}(S) \right)$$

$$= (\delta_{m_n}(S) - B) + \frac{3 - \delta_{m_n}(S)}{m_n + 1}$$

$$\leq \epsilon.$$

This proves the first statement.

Now suppose $p = 2$. Then $\delta_{m_n}(S) < 1/2$ implies that

$$\frac{3}{2} \Phi_{m_n}(S) < \frac{3 \cdot 2^{m_n}}{2\Phi_{m_n}(S)} \quad \text{and} \quad \sqrt{6} \cdot 2^{m_n/2} < \frac{\sqrt{6} \cdot 2^{m_n}}{\Phi_{m_n}(S)}.$$

Thus by Lemma 3.2 we have

$$\Phi_{m_n+1}(S) < \frac{(3 + \sqrt{6})2^{m_n}}{\Phi_{m_n}(S)}.$$

Thus

$$\lambda_{m_n+1}(S) < m_n + \log_2(3 + \sqrt{6}) - \lambda_{m_n}(S),$$

so

$$\delta_{m_n+1}(S) - (1 - B) \leq \frac{m_n + \log_2(3 + \sqrt{6})}{m_n + 1} - \frac{m_n}{m_n + 1} \delta_{m_n}(S) - (1 - B)$$

$$< (B - \delta_{m_n}(S)) + \frac{\log_2(9/2) + \delta_{m_n}(S)}{m_n + 1}$$

$$\leq \epsilon.$$

Thus $|1 - B - \delta_{m_n+1}(S)| < \epsilon$ for n sufficiently large, proving the lemma.   □

Corollary 1. *In general $1 - B \leq C$ and $1/2 \leq C$.*

Proof: By Lemma 4, if $B < 1/2$, then $1 - B > 1/2$ is an accumulation point. If $B \geq 1/2$, then $C \geq B \geq 1/2 \geq 1 - B$. In either case $C \geq 1 - B$ and $C \geq 1/2$.   □

We can now prove our main result.

Theorem 3. *Let S be a binary sequence (i.e., $p = 2$) and suppose that the set of accumulation points of the set of $\delta_n(S)$ is the interval $[B, C] = \{a \in \mathbb{R} : B \le a \le V\}$. Then $B \le \log_2(3/2)$. If $B < \log_2(4/3)$ then $C = 1 - B$. That is, $T_\pi(S) = [B, 1 - B]$.*

Proof: There is a sequence of integers $\ell_1 < \ell_2 < \cdots$ such that $\lim_{n \to \infty} \delta_{\ell_n}(S) = C$ and we can assume that $|C - \delta_{\ell_n}(S)| > |C - \delta_{\ell_{n+1}}(S)|$ for all n. By possibly deleting some of the ℓ_n and m_n, we can assume that $m_n < \ell_n < m_{n+1}$ for all $n \ge 1$. For n sufficiently large we have $\delta_{m_n} < \delta_{\ell_n}$, so we can assume this holds for all $n \ge 1$. Thus there is an $\ell \le \ell_n$ so that $\delta_{\ell-1}(S) < \delta_\ell(S)$. If we replace ℓ_n by the largest such ℓ, then we still have a sequence whose limit is C. So we can assume that $\delta_{\ell_n-1}(S) < \delta_{\ell_n}(S)$ for all n. In particular, $\Phi_{\ell_n-1}(S) < \Phi_{\ell_n}(S)$. Then by Lemma 3.3, for every $\epsilon > 0$ if n is sufficiently large, then

$$\delta_{\ell_n-1}(S) < \max(1/2 + \epsilon, \log_2(3(1 + 2\epsilon)/2)). \tag{8}$$

This implies that there is an accumulation point of the $\delta_n(S)$ that is less than or equal to $\max(1/2, \log_2(3/2)) = \log_2(3/2)$, so $B \le \log_2(3/2)$. This proves the first statement.

To prove the second statement, let us assume to the contrary that $1 - B < C$ and that $B \le \log_2(4/3) < 1/2$. Thus $C > 1 - \log_2(4/3) = \log_2(3/2)$. By part (1) of Lemma 3,

$$\Phi_{\ell_n}(S) \le \frac{3\Phi_{\ell_n-1}(S)}{2} + \frac{2^{\ell_n-1}}{\Phi_{\ell_n-1}(S)} + \sqrt{6} \cdot 2^{(\ell_n-1)/2} \le 3\max\left(\frac{3\Phi_{\ell_n-1}(S)}{2}, \frac{2^{\ell_n-1}}{\Phi_{\ell_n-1}(S)}, \sqrt{3} \cdot 2^{\ell_n/2}\right).$$

Thus

$$\delta_{\ell_n}(S) \le \max\left(\frac{\ell_n - 1}{\ell_n}\delta_{\ell_n-1}(S) + \frac{\log_2(9/2)}{\ell_n}, 1 - \frac{\ell_n - 1}{\ell_n}\delta_{\ell_n-1}(S) + \frac{\log_2(3/2)}{\ell_n},\right.$$
$$\left.\frac{1}{2} + \frac{\log_2(3\sqrt{3})}{\ell_n}\right). \tag{9}$$

There are three cases to consider, depending on which term is maximal. Suppose that

$$\delta_{\ell_n-1}(S) \le \frac{1}{2} - \frac{\log_2(6)}{2(\ell_n - 1)}.$$

Then the right hand side of equation (9) equals the second term, so

$$\delta_{\ell_n-1}(S) \le \frac{\ell_n}{\ell_n - 1} - \frac{\ell_n}{\ell_n - 1}\delta_{\ell_n}(S) + \frac{\log_2(3/2)}{\ell_n - 1}.$$

If this occurs for infinitely many n, then the set $\{\delta_{\ell_n-1}(S) : n \ge 1\}$ has an accumulation point less than or equal to

$$\lim_{n \to \infty} \frac{\ell_n}{\ell_n - 1} - \frac{\ell_n}{\ell_n - 1}\delta_{\ell_n}(S) + \frac{\log_2(3/2)}{\ell_n - 1} = 1 - \lim_{n \to \infty} \delta_{\ell_n}(S) = 1 - C < B.$$

This is a contradiction, so (by possibly deleting finitely many ℓ_ns) we may assume that $\delta_{\ell_n-1}(S) > 1/2 - \sqrt{6}/(\ell_n - 1)$ for every n.

Suppose that

$$\frac{1}{2} - \frac{\log_2(6)}{2(\ell_n - 1)} < \delta_{\ell_n-1}(S) \leq \frac{1}{2} + \frac{\log_2(12)}{2(\ell_n - 1)}.$$

Then the right hand side of equation (9) equals the third term, so

$$\delta_{\ell_n}(S) \leq \frac{1}{2} + \frac{\log_2(3\sqrt{3})}{\ell_n}.$$

If this occurs for infinitely many n, then

$$C = \lim_{n\to\infty} \delta_{\ell_n}(S) \leq \lim_{n\to\infty} \frac{1}{2} + \frac{\log_2(3\sqrt{3})}{\ell_n} = \frac{1}{2} < C.$$

This is a contradiction, so (by possibly deleting finitely many ℓ_ns) we may assume that

$$\delta_{\ell_n-1}(S) > \frac{1}{2} + \frac{\log_2(12)}{2(\ell_n - 1)}$$

for every n.

Thus the right hand side of equation (9) equals the first term, and

$$C = \lim_{n\to\infty} \delta_{\ell_n}(S) \leq \lim_{n\to\infty} \frac{\ell_n - 1}{\ell_n} \delta_{\ell_n-1}(S) + \frac{\log_2(9/2)}{\ell_n} = \lim_{n\to\infty} \delta_{\ell_n-1}(S).$$

But C is the maximum accumulation point of the $\delta_i(S)$, so in fact $\lim_{n\to\infty} \delta_{\ell_n-1}(S) = C$.

It then follows from equation (8) that $C \leq \max(1/2, \log_2(3/2)) = \log_2(3/2)$, which is a contradiction. $\qquad\square$

4   Existence Results for $T_\pi(S)$s

In this section we return to the general setting where $p \geq 2$ is a square free positive integer and $\pi^2 = -p$. Thus we are now considering sequences over the more general set $\{0, 1, \cdots, p - 1\}$, and we are considering π-adic complexity based on the more general ring $R = \mathbb{Z}[\sqrt{-p}]$. We denote by β_π the function that associates the real number B with the sequence S, where $T_\pi(S) = [B, C]$. That is,

$$\beta : \{S = s_0, s_1, \cdots, s_i \in \{0, 1\}\} \to [0, 1]$$

and $\beta_\pi(S)$ is the least accumulation points of the set of normalized π-adic complexities of prefixes of S. In this section we see that the image of β contains $[0, 1/2]$.

Theorem 4. *For every $p \geq 2$ and every $B \in [0, 1/2]$ there is a p-ary sequence S with $T_\pi(S) = [B, C]$ for some C. If $p = 2$, then we can take $C = 1 - B$.*

Proof: Let $0 \leq B < 1/2$. We build S with $\beta(S) = B$ in stages. Suppose that we have chosen $n_1 \leq n_2 \leq \cdots \leq n_r \in \mathbb{Z}^+$ and $S^{n_r} = s_0, \cdots, s_{n_r-1} \in \{0, 1, \cdots, p-1\}$ so that

$$|\delta_{n_i}(S) - B| \leq \frac{1}{n_i},$$

$\delta_{n_i}(S) \leq B < \delta_j(S)$ for $n_i < j < n_{i+1}$, and

$$n_1 > \max\left(\frac{1 + \log_p(4)}{1 - 2B}, \frac{B+1}{1 - 2B}\right). \tag{10}$$

Choose s_{n_r} so that $\Phi_{n_r+1}(S) \neq \Phi_{n_r}(S)$. Some care must be taken here to see that this is in fact possible. Suppose that a/b, $a, b \in R$ with π invertible modulo b, is the rational approximation to $\sum_{i=0}^{r-1} s_i\pi^i$ modulo π^{n_r} so that $\Phi(a, b)$ is minimal. Suppose also that $a/b \equiv \sum_{i=0}^{n_r-1} s_i\pi^i + s'_{n_r}\pi^{n_r} \pmod{\pi^{n_r+1}}$. We choose $s_{n_r} \neq s'_{n_r}$. Then the bound in equation (10) and the fact that $\Phi(a, b) < p^{n_r B+1}$ ensure that there are not integers c and d so that $c/d \equiv \sum_{i=0}^{n_r} s_i\pi^i \pmod{\pi^{n_r+1}}$ with $\Phi(a, b) = \Phi(c, d)$. That is, the π-adic complexity profile must increase at this point. By Lemma 3.1,

$$\delta_{n_r+1}(S) > \frac{n_r - 1}{n_r + 1} - \frac{n_r}{n_r + 1}\delta_{n_r}(S)$$

$$\geq \frac{n_r - 1}{n_r + 1} - \frac{n_r}{n_r + 1}B$$

$$> B,$$

where the last inequality follows from equation (10). As in the proof of Lemma 4, the limit of the δ_{n_r+1} is at least $1 - B$. If $p = 2$, then the limit of the δ_{n_r+1} is exactly $1 - B$.

Now choose n_{r+1} and $s_{n_r+1}, \cdots, s_{n_{r+1}-1}$ so that $\Phi_{n_r+1}(S) = \cdots = \Phi_{n_{r+1}}(S)$ and $\delta_{n_{r+1}}(S) \leq B < \delta_{n_{r+1}-1}(S)$. This is possible since the $\delta_j(S)$ are decreasing with limit 0 if the $\Phi_j(S)$ are unchanged. Moreover, for any j such that $\Phi_{j+1}(S) = \Phi_j(S)$, we have $\lambda_{j+1}(S) = \lambda_j(S)$ and

$$\delta_j(S) - \delta_{j+1}(S) = \frac{\lambda_j(S)}{j} - \frac{\lambda_j(S)}{j+1}$$

$$= \frac{\lambda_j(S)}{j(j+1)}$$

$$\leq \frac{1}{j+1}.$$

It follows that $B - 1/n_{r+1} < \delta_{n_{r+1}}(S) < B$. Thus $B = \lim_{i \to \infty} \delta_{n_i}(S)$. It also follows that B is the least accumulation point of the $\delta_j(S)$. Also, as in the proof of Lemma 4, the limit of the δ_{n_r+1} is at least $1 - B$. If $p = 2$ the limit equals $1 - B$, the maximum accumulation point.

Finally, let $B = 1/2$. We construct S a term at a time as follows. If $\delta_r(S) < 1/2 - 2/r$, then choose s_r so that $\Phi_r(S) \neq \Phi_{r+1}(S)$. As before, we have made

$\delta_r(S)$ small enough that $\Phi_r(S) \neq \Phi_{r+1}(S)$ if we choose the "wrong" s_r. If $\delta_r(S) \geq 1/2 - 2/r$, then choose s_r so that $\Phi_r(S) = \Phi_{r+1}(S)$. We claim that $\lim_{r\to\infty} \delta_r(S) = 1/2$, so that $1/2$ is the only accumulation point.

Let r be an arbitrary index. If $\delta_r(S) \geq 1/2 - 2/r$, then $\delta_{r+1}(S) \geq 1/2 - 3/(2(r+1))$, which guarantees that the least accumulation point is at least $1/2$. If $\delta_r(S) < 1/2 - 2/r$, then

$$\delta_{r+1}(S) > \frac{r-1}{r+1} - \frac{r}{r+1}\delta_r(S)$$
$$> \frac{r-1}{r+1} - \frac{r}{r+1}\left(\frac{1}{2} - \frac{2}{r}\right)$$
$$= \frac{1}{2} + \frac{1}{2(r+1)}.$$

On the other hand, suppose $\delta_r(S) > 1/2 + 1/r$. Then

$$\delta_{r+1}(S) = \frac{r}{r+1}\delta_r(S) > \frac{1}{2} + \frac{1}{2(r+1)}.$$

If we apply our constructions and $\delta_j(S) > 1/2 - 2/j$ for $j = r, r+1, \cdots, k$ for some k, then

$$\delta_k(S) = \frac{r}{k}\delta_r(S).$$

Thus after finitely many steps we reach a k for which $\delta_k(S) < 1/2 - 2/k$. (In fact the first k for which $\delta_k(S) \leq 1/2$ is at most $k = 2(r+2)$.) We have shown that $\delta_r(S) \in [1/2 - 1/r, 1/2]$ for infinitely many r, and that $\delta_r(S) > 1/2 - 1/r$ for all r. Thus $B = 1/2$ is the least accumulation point. Again, as in the proof of Lemma 4, if $p = 2$, then the limit of the δ_{r+1} for which $\Phi_r(S) \neq \Phi_{r+1}(S)$ is $1 - 1/2 = 1/2$ and this is the maximum accumulation point. This completes the proof.   □

Note that if we modify this construction so that the the N-adic complexity changes when $\delta_{n_r}(S) < \delta_{n_r-1}(S) < B < \delta_{n_r-2}(S)$, then B is still the least accumulation point. In fact, at each phase we can either use this method or the one in the proof to determine when to change the N-adic complexity. Since there are infinitely many phases, this gives uncountably many sequences for which B is the least accumulation point.

Corollary 2. *For any B with $0 \leq B \leq 1/2$, there are uncountably many sequences S with $\beta(S) = B$. If $p = 2$, then there are uncountably many sequences S with $T_\pi(S) = [B, 1 - B]$.*

5   Conclusions

We have found constraints on the possible sets of accumulation points of the normalized π-adic complexities of non-periodic sequences. This gives us a fuller understanding of the properties of these security measures. It provides another in a growing list of ways that feedback with carry shift registers are similar to linear feedback shift registers, although here we see that precise analysis depends on strong number theoretic properties of the ring R. There are also practical implications to this and earlier work on linear and N-adic complexity. Suppose a stream cipher uses an infinite non-periodic sequence S as a keystream (of course one may question whether such a keystream possible since no finite state device can output a nonperiodic sequence), and suppose that the set of accumulation points of the normalized π-adic or linear complexity is $[B, 1 - B]$. Now imagine a cryptanalyst who has observed a prefix of S and wants to predict the next symbol. If the normalized complexity up to this point is close to B, the next symbol is likely to change the complexity so the normalized complexity increases. Likewise, if the normalized complexity up to this point is close to $1 - B$, then the next symbol is likely to leave the complexity unchanged so the normalized complexity decreases. In this sense sequences for which the set of accumulation points is $[0, 1]$ are the most random sequences.

We would like to know whether $T(S)$ is of the form $[B, 1 - B]$ for all p and all S. Even this is just a start — the same questions can be asked for every setting R, π for which we have algebraic feedback shift registers. Moreover, as with linear complexity we can consider the asymptotic π-adic complexity of multi-sequences.

References

1. Borevich, Z.I., Shefarevich, I.R.: Number Theory. Academic Press, New York (1966)
2. Dai, Z., Imamura, K., Yang, J.: Asymptotic behavior of normalized linear complexity of multi-sequences. In: Helleseth, T., Sarwate, D., Song, H.-Y., Yang, K. (eds.) SETA 2004. LNCS, vol. 3486, pp. 126–142. Springer, Heidelberg (2005)
3. Dai, Z., Jiang, S., Imamura, K., Gong, G.: Asymptotic behavior of normalized linear complexity of ultimately non-periodic sequences. In: IEEE Trans. Info. Theory, vol. 50, pp. 2911–2915. IEEE Computer Society Press, Los Alamitos
4. Feng, X., Dai, X.: The expected value of the normalized linear complexity of 2-dimensional binary sequences. In: Helleseth, T., Sarwate, D., Song, H.-Y., Yang, K. (eds.) SETA 2004. LNCS, vol. 3486, pp. 113–128. Springer, Heidelberg (2005)
5. Klapper, A.: The Asymptotic Behavior of 2-Adic Complexity, under review
6. Klapper, A.: Distribution properties of d-FCSR sequences. Journal of Complexity 20, 305–317 (2004)
7. Klapper, A., Goresky, M.: Feedback Shift Registers, 2-Adic Span, and Combiners with Memory. J. Cryptology 10, 111–147 (1997)
8. Klapper, A., Xu, J.: Algebraic feedback shift registers. Theoretical Comp. Sci. 226, 61–93 (1999)
9. Klapper, A., Xu, J.: Register synthesis for algebraic feedback shift registers based on non-primes. Designs, Codes, and Cryptography 31, 225–227 (2004)
10. Massey, J.: Shift-register synthesis and BCH decoding. IEEE Trans. Infor. Theory IT-15, 122–127 (1969)
11. Niederreiter, H.: The probabilistic theory of linear complexity. In: Günther, C.G. (ed.) EUROCRYPT 1988. LNCS, vol. 330, pp. 191–209. Springer, Heidelberg (1988)

Shannon Capacity Limits of Wireless Networks

Andrew Viterbi

Viterbi Group, LLC

Abstract. Two wideband physical-layer multiple access techniques for mobile telephony and data are compared: CDMA, which in one of several manifestations has been chosen for virtually all third generation cellular systems, and OFDM with MIMO, which seems to be the most favored for a future generation. We compare the Shannon capacities of the two and conclude that with appropriate processing CDMA and OFDM capacities are similar but that the latter may more effectively be combined with MIMO antennas to provide higher capacities.

1 Summary

The physical layer of wireless networks has evolved through three generations in less than three decades. As we approach the fourth decade of cellular wireless service, yet a fourth generation is being proposed. Table 1 summarizes the technologies adopted for each generation whose evolution lasted approximately one decade between initial deployment and maturity at which time a new generation arose. Currently advanced versions of Code Division Multiple Access (CDMA-EVDO and W-CDMA) are reaching maturity. While through further improvements these are likely to retain major market shares for the next decade, an alternate technology, orthogonal frequency division multiplexing (OFDM), also known by its trade name Wi-MAX, is taking hold. Along with multiple antennas, referred to as multiple-in multiple-out (MIMO) systems, this seems to be favored for future deployments. Table 1 also provides an estimate of the bandwidth efficiency of each generation. The last entry depends strongly on the success of MIMO to provide a capacity enhancement nearly proportional to the number of antenna pairs, which will be discussed in the last section. In the following three sections we consider the capacity of single- versus multiple-carrier systems, first in multipath-fading and then in the presence of multiple-user interference. We then turn to multiple antennas and consider their possible enhancement of both types of systems.

Table 1. Cellular Generations

Decade	Generation		Efficiency: bps/Hz/sector
1980's	1G	Analog Cellular	.016
1990's	2G	Digital (TDMA→CDMA)	.05 → .2
2000's	3G	Enhanced CDMA	.4 → .6
2010's	4G	OFDM/MIMO	> 1.0

S.W. Golomb et al. (Eds.): SSC 2007, LNCS 4893, pp. 147–152, 2007.

All comparisons throughout will be in terms of the Shannon capacity of each system, as justified by the following three considerations:

a) It represents a hard upper limit on throughput performance;
b) This limit is almost achievable, if we allow for some processing delay, by LDPC and Turbo codes;
c) For Gaussian interference, a reasonable model for wideband channels, the capacity expression is simple: $C = W \log(1 + S/I)$, where C is capacity in bits/sec., W is bandwidth in Hz and S/I is signal-to-interference ratio.

2   Capacities in Multipath-Fading Channels

Figure 1 shows the models for single-carrier and multiple-carrier channels. The time model better suits the former while the frequency model better suits the latter. An example of a single-carrier system is CDMA, while OFDM is the embodiment of a multiple carrier system. In all cases, we may assume that the channel parameters are known at the receiver through measurements on pilot signals. The optimum receiver for the time model is a matched filter (or rake receiver), while for the frequency model it is a bank of receivers each tuned to the appropriate frequency and appropriately scaled. The tap-coefficients h_j are related to the frequency gains k_n by the discrete Fourier transform

$$k_n = \sum_{j=0}^{L-1} h_j \exp(-2\pi i n j / L). \tag{1}$$

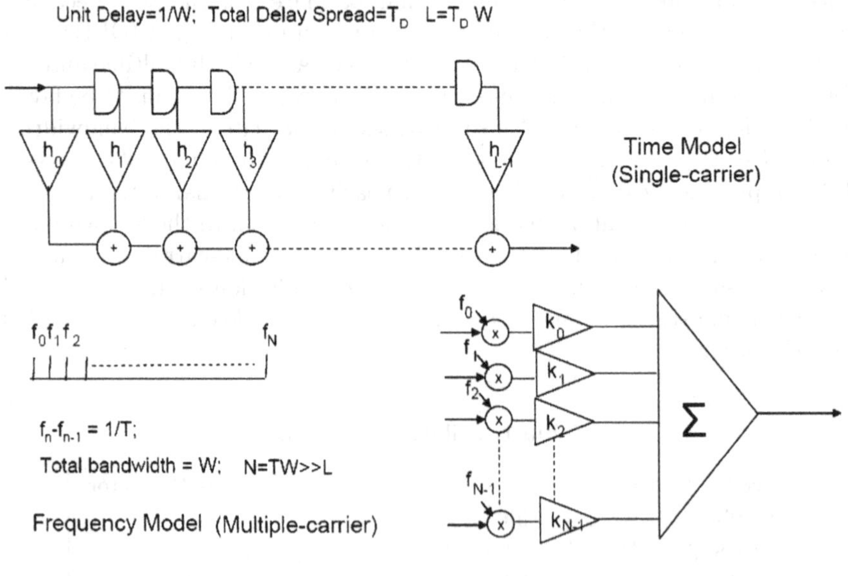

Fig. 1. Multipath-Fading Channel

The power relations between the two sets follow (provided N is a multiple of L, an easily accommodated condition):

$$\sum_{n=0}^{N-1} |k_n|^2/N = \sum_{j=0}^{L-1} |h_j|^2, \tag{2}$$

which is just the discrete form of Parseval's Theorem. From this and Jensen's Inequality for convex functions, we may compare the capacities of single- and multiple-carrier systems, which we label C_C and C_O (for CDMA and OFDM respectively),

$$C_C = W \log \left[1 + (S/I) \sum_{j=0}^{L-1} |h_j|^2 \right], \tag{3}$$

while using Jensen's Inequality and Parseval's Theorem,

$$C_O = (W/N) \sum_{n=0}^{N-1} \log \left[1 + (S/I)|k_n|^2 \right]$$

$$\leq W \log \left[1 + (S/I) \sum_{n=0}^{N-1} |k_n|^2/N \right] = C_C. \tag{4}$$

Note, however, that for small values of S/I the logarithms approach linear functions and hence the inequality approaches an equality.

While this appears to show an advantage for CDMA over OFDM, the tables appear to be turned when other-user interference is considered.

3   Capacities in the Presence of Other-User Interference

Without loss of generality we may let the total powers (2) in the tap coefficients (or in the normalized frequency multipliers) be set at unity, tantamount to considering them part of the received signal power. Assume further for the sake of simple comparison that all user signals are power controlled so that they arrive at the base station with equal powers S and that the user population is uniformly distributed with M users occupying each cell (or sector). Then CDMA in which each user occupies the total spectrum will have $M - 1$ interfering users in its own cell and M users in all other cells. It has been shown [1] that because of the increased distances from the base station, the totality of other-cell interference is equivalent to the power from ρM users in the given cell, where depending on the propagation law, ρ lies between 0.6 and 1.0 [1]. Consequently, since all same-cell users are power controlled to the same level, the total interference in CDMA is $I = (M-1)S + \rho MS \sim (1+\rho)MS$. Thus for CDMA, $S/I \sim 1/[(1+\rho)M]$. Since this is very small, we may linearize the logarithm so that the per-user capacity becomes $W/[(1+\rho)M \ln(2)]$. The total throughput per cell (or sector) for CDMA for all M users becomes M times this,

$$C_{C\ Total} \sim W/[(1 + \rho) \ln(2)]. \tag{5}$$

For OFDM, on the other hand, each same-cell user can be assigned a unique set of carriers (changed on each successive symbol), so that only other-user interference affects each user. In that case, each user occupies only W/M Hz and is affected by $1/M$th of the interference, so $S/I = 1/\rho$ and the total OFDM throughput per cell (or sector) for all M users becomes

$$C_{O\ Total} \sim W \log[1 + (1/\rho)]. \tag{6}$$

Consequently, for values of ρ in the range of 0.6 to 1.0, OFDM capacity exceeds that of CDMA. We next show how CDMA may be redeemed by successive interference cancellation.

4   CDMA with Successive Interference Cancellation

By a process of successive cancellation, CDMA can achieve the same total throughput capacity as OFDM. Suppose the Mth user is demodulated and decoded. As long as its rate is below its capacity as determined by its S/I, the result should be correct. Then if it is re-encoded, remodulated and subtracted from the composite received signal, its effect will be removed from the interference to the other users. This is then repeated $M - 2$ more times with the result that the mth user will only be affected by $m - 1$ same-cell user interference along with that of all users in all other cells. We take the total other-cell interference as before to be ρMS, so the interference seen by the mth user will be $(m - 1)S + \rho MS$.

Then letting individual rates approach capacity, the rate for the mth user will be

$$R_m/W = \log\{1 + S/[\rho MS + (m - 1)S]\} = \log\{[\rho M + m]/[\rho M + (m - 1)]\}.$$

Since the sum of logs equals the log of the product and all successive terms of the product will cancel except the numerator of the first and the denominator of the last, the total throughput per cell (sector) becomes

$$C_C = \sum_{m=1}^{M} R_m = W \log[(\rho M + M)/\rho M] = W \log[1 + (1/\rho)] = C_O. \tag{7}$$

The drawback, however, is that the rates are all different with the Mth user's smallest and the rest growing as m decreases. All rates can be made equal by varying signal powers, but then these will grow almost exponentially. A technique which achieves equal rates with equal powers is illustrated for $M = 3$ users in Figure 2. Here the users are staggered in time and decoded and cancelled in the order shown. Since 1/3 of the time they see only other-cell interference, 1/3 one other user along with other-cell and 1/3 two other users along with other-cell interference, all users achieve the same rate with the same power.

Stagger User Packets (example for 3 users)

1	2	3	10	11	12		
4	5	6	13	14	15		
7	8	9	16	17	18		

$$C_m = (W/3) \left[\log(1 + S/I) + \log(1 + \frac{S/I}{1 + S/I}) + \log(1 + \frac{S/I}{1 + 2S/I}) \right]$$
$$= (W/3) \log(1 + 3S/I); \; m = 1, 2, 3$$
$$C_{Total} = W \log(1 + 3S/I)$$

Fig. 2. Equal Rates with Equal Powers

Then extending to M staggered users, with $I = MS\rho$, the expression for total throughput capacity in Figure 2 generalizes to:

$$C_C = \sum_{m=1}^{M} R_m = W \log[1 + MS/I] = W \log[1 + (1/\rho)] = C_O. \qquad (8)$$

5   Multiple-Input Multiple-Out (MIMO) Antenna Systems

The added dimensions provided by MIMO antenna systems can be employed to provide redundancy to mitigate fading or to increase data rates, or a combination of both. Since our focus here is on capacity, we consider only the increased rate capability. For simplicity of discussion we shall take the number of transmit and receive antennas to be equal to N_A, though this is not necessary in general. Letting x be the N_A inputs and y the N_A outputs, the input-output relations are given by the vector equation

$$y = Hx + w,$$

where H is the $N_A \times N_A$ transfer matrix of channel gains and w is the vector of noises at the receiver. Employing the Singular Value Decomposition theorem, H can be written as

$$H = U\Lambda V^*,$$

where U and V are unitary matrices (rotations): $UU^* = I = VV^*$, and Λ is a diagonal matrix whose components are the eigenvalues of H.

Figure 3 illustrates the above. The model does however need knowledge of channel state information (CSI) at the transmitter as well as at the receiver (where it is usually measured). With such knowledge, the relative signal powers S_n can be chosen to maximize capacity based on the common "water-filling"

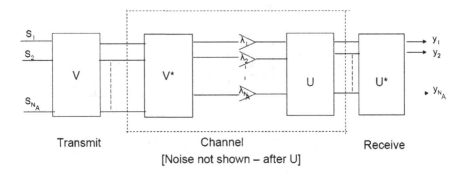

$$C = W \sum_{n=1}^{N_A} \log(1 + \frac{S_n \lambda_n^2}{I})$$

Fig. 3. Capacity of MIMO with CSI

technique. Without such information, the transmit rotator V will be omitted, but if we let all antenna inputs be equal to S/N_A, where S is the total power and N_A is the number of antenna-pairs, since the rotator does not change the power levels, the capacity expression is fundamentally unchanged:

$$C = W \sum_{n=1}^{N_A} \log[1 + \frac{S \lambda_n^2}{I N_A}]. \tag{9}$$

Depending on the values of the eigenvalues, the capacity may increase proportional to the number of antenna-pairs. Note, however, that in a single-carrier system the H matrix pertains to propagation over the entire wide bandwidth which may result in rapid variations. On the other hand, in a multiple-carrier system each subcarrier will have its own unique H matrix which will be much more stable and the receiver can thus operate on each subcarrier individually. This is a distinct advantage for OFDM over CDMA.

Summarizing, the reduced interference advantage of OFDM can be equalized in CDMA by successive interference cancellation. Only when MIMO is employed may OFDM have a distinct advantage over CDMA systems.

Acknowledgement

The author is grateful to Prof. J.K. Wolf for the concepts presented in Sections 3 and 4.

References

1. Viterbi, A.J.: CDMA: Principles of Spread Spectrum Communication. Addison-Wesley, Reading (1995)

Some Mysterious Sequences Associated with LDPC Codes[*]

Robert J. McEliece and Sarah L. Sweatlock

California Institute of Technology

1 Introduction

One of the most important research areas in coding theory is *weight enumeration*. This is a large subject, but the basic problem is easily stated: determine or estimate the *weight enumerator* $(B_0, \ldots, B_n)$ for an (n, k) binary linear code, specified by a $(n - k) \times n$ parity-check matrix $\mathcal{H}$ with entries from $GF(2)$. Here

$$B_i = \#\{c \in GF(2)^n : \mathcal{H}c^T = 0, \text{wt}(c = i)\},$$

where $\text{wt}(c)$ is the weight of the vector c. If the number of codewords is large, the logarithmic weight enumerator, i.e.,

$$(\frac{1}{n} \log B_0, \ldots, \frac{1}{n} \log B_n)$$

is more convenient. If a code belongs to a family of codes which share similar properties, the log-weight enumerator may approach a limiting function called the *spectral shape*:

$$\frac{1}{n} \log \left(B_{\lfloor \theta n \rfloor} \right) \to E(\theta), \qquad 0 < \theta < 1.$$

In modern coding theory, $\mathcal{H}$ is usually very large and very sparse, e.g., the row and column sums are bounded as $n \to \infty$. The corresponding codes are called *low density parity-check codes*. Often we are faced with large collections, or *ensembles*, of long LDPC codes, which share similar properties, in which case it may be difficult to find the spectral shape of an individual member of the ensemble, but relatively easy to calculate the *ensemble average*.

In this paper, we study two popular LDPC ensembles defined by restricting the number of ones in each row and/or column of $\mathcal{H}$: the (j, k) ensemble and the $(*, k)$ ensemble. The (j, k) ensemble, introduced by Gallager [1] and included in a later study by Litsyn and Shevelev [2] consists, roughly speaking, of the binary codes represented by parity check matrices with j ones in each column and k ones in each row. The spectral shape of the (j, k) ensemble is given by Equations 1-5, [1],[2]. The $(*, k)$ ensemble, which was not considered by Gallager, but was included in the study by Litsyn and Shevelev, consists of the binary codes represented by parity check matrices with k ones in each row but no restrictions

---

[*] This research was supported by the Lee Center for Advanced Networking and the Sony Corporation.

S.W. Golomb et al. (Eds.): SSC 2007, LNCS 4893, pp. 153–161, 2007.
© Springer-Verlag Berlin Heidelberg 2007

on the columns. The spectral shape of the $(*, k)$ enumerator was determined in [2] and is given in Equation 6.

The Entropy Function: $\qquad H(x) = -x \log x - (1-x) \log(1-x) \qquad (1)$

The Partition Function: $\qquad Z(s) = \dfrac{1}{2} \left((1-e^{-s})^k + (1+e^{-s})^k \right) \qquad (2)$

The Free Energy: $\qquad f(s) = -\log Z(s) \qquad (3)$

$$\theta(s) = \frac{f'(s)}{k} \qquad (4)$$

$$E(s) = \frac{j}{k} \left(sf'(s) - f(s) \right) - (j-1)H(\theta(s)). \quad (5)$$

$$E(\theta) = H(\theta) + (1-R) \log \left(\frac{1 + (1-2\theta)^k}{2} \right), \qquad (6)$$

where R is the rate of the code.

Note: Litsyn and Shevelev considered 8 different ensembles of LDPC codes labeled A-H. Three of these ensembles, viz. A,B and H, have identical spectral shapes, though the ensembles are slightly differerent. Indeed, Gallager's ensemble is Litsyn and Shevelev's ensemble B. The ensemble we are calling $(*, k)$ has the same spectral shape as ensembles E and F.

The spectral shapes for various rate $1/2$ codes are shown in Figures 1 and 2. Figure 1 shows the (j, k) ensembles, and Figure 2 shows the $(*, k)$ ensembles. Both are shown along with the shifted entropy function $h_s(\theta, R) = H(\theta) - (1-R) \log(2)$. Note that each of the spectral shapes is tangent to the shifted entropy function at $\theta = \frac{1}{2}$. We are interested in the degree of tangency and more generally in the power series expansion of the spectral shape centered at that point. However, calculating the coefficients of the power series expansion is not a simple task.

The organization of this paper is as follows: in Section 2 we supply some techniques which simplify the calculation of the power series expansion. The reader who is not interested in the mathematics should skip to Sections 3 and 4 to see the results. Section 5 supplies a brief conclusion and discussion of future work.

2 Exponential Generating Functions

Note: This section is based on Chapter 5 of Richard Stanley's two-volume study of Enumerative Combinatorics [3]. The reader already familiar with Stanley's work will learn little in this section.

A *sequence* is a mapping from the nonnegative integers $\mathbb{N}$, or the positive integers $\mathbb{P}$, to a fixed field K of characteristic zero. Typically, such a sequence will be denoted by $f(0), f(1), f(2) \ldots$. The *exponential generating function* for the sequence $f(n)$ is the formal power series

$$E_f(x) = \sum_{n \geq 0} f(n) \frac{x^n}{n!}.$$

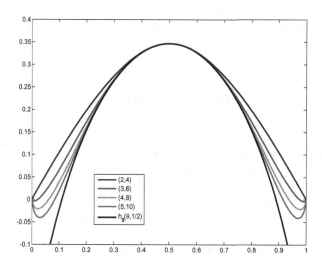

Fig. 1. Spectral Shapes of Rate $1/2$ (j,k) Ensembles

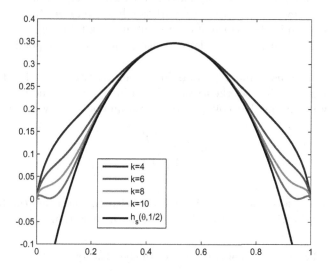

Fig. 2. Spectral Shapes of Rate $1/2$ $(*,k)$ Ensemble

Note that the original *sequence* $f(n)$ can be recovered from the *series* $E_f(x)$ by formal differentiation:

$$f(n) = \left.\frac{d^n}{dx^n} E_f(x)\right|_{x=0}$$

We summarize this important pairing with the symbol

$$f(n) \leftrightarrow E_f(x). \tag{7}$$

The problems we shall address in this section are as follows. Given sequences $f : \mathbb{P} \to K$, with $f(1) = 1$, and $g : \mathbb{N} \to K$, with $g(0) = 1$, construct sequences $k(n)$, $h_0(n)$, and $h(n)$ such that

$$E_k(x) = E_g(E_f(x)) \tag{8}$$

$$E_{h_0}(x) = E_f^{(-1)}(x) \tag{9}$$

$$E_h(x) = E_g(E_f^{(-1)}(x)) \tag{10}$$

These constructions are notationally inconvenient but not conceptually difficult. The basic notion is that of a partition of an integer n, i.e., a multiset of positive integers that sums to n. Thus $\{\{1, 1, 2, 2, 2\}\}$ is a partition of 8. Functions of λ are defined to be simply the product of the functions of each λ_i, ie, if $\lambda = \{\{\lambda_1, \lambda_2, \ldots, \lambda_k\}\}$, then for a generic f, $f(\lambda) = f(\lambda_1)f(\lambda_2) \ldots f(\lambda_k)$. A partition of n is also characterized by its multiplicities $m_j = \# \{i : \lambda_i = j\}$: $n = 1^{m_1} 2^{m_2} \cdots$. Thus, the partition $\{\{1, 1, 2, 2, 2\}\} = 1^2 2^3$, and in general, $f(\lambda) = \prod_j f(j)^{m_j}$.

We will also need the notion of *set theory* partitions. Every set theory partition has a corresponding number theory partition, which is simply a list of the size of each subset, and every number theory partition has several set theory partitions that correspond to it. As a convention, we will use capital letters, like Λ, for set theory partitions and lower-case letters, like λ, for the corresponding number theory partition.

Let $\#\lambda$ denote the number of parts in a given partition λ and r denote the number of singletons in a set theory partition. Finally, $N(\lambda)$ denotes the number of set theory partitions that collapse to a given number theory partition.

$$N(\lambda) = \frac{n!}{\displaystyle\prod_{i=1}^{n}(i!)^{m_i} m_i!}$$

Table 1 lists a few examples, with $n = 5$. With this notation established, it should be easy for the reader to follow the next three sections.

Table 1. Examples of Set and Number Theory Partitions

Set Theory Partition Λ	Number Theory Partition λ	$\#\lambda$	r	$N(\lambda)$
(12)(3)(45)	$\{1, 2, 2\}$	3	1	15
(13)(4)(25)	$\{1, 2, 2\}$	3	1	15
(15)(24)(3)	$\{1, 2, 2\}$	3	1	15
(12345)	$\{5\}$	1	0	1
(12)(345)	$\{2, 3\}$	2	0	10
(123)(45)	$\{2, 3\}$	2	0	10
(1234)(5)	$\{1, 4\}$	2	1	5

2.1 Faà di Bruno's Problem

Theorem 1. *(Faà di Bruno's Theorem)*
Given two sequences $f(n)$ and $g(n)$, let $k(n)$ be the sequence defined by $E_k(x) = E_g(E_f(x))$ (see equation 7). Then,

$$k(n) = \sum_{\lambda \in P(n)} N(\lambda) f(\lambda) g(\#\lambda) \text{ where}$$

$$P(n) = \left\{ \lambda : \sum_j \lambda_j = n \right\}$$

Example $P(n)$'s:
$$P(1) = \{1\}$$
$$P(2) = \{2, 11\}$$
$$P(3) = \{3, 21, 111\}$$
$$P(4) = \{4, 31, 22, 211, 1111\}$$

Example $k(n)$'s:
$$k(0) = 1$$
$$k(1) = g(1)$$
$$k(2) = f(2)g(1) + g(2)$$
$$k(3) = f(3)g(1) + 3f(2)g(2) + g(3)$$
$$k(4) = f(4)g(1) + 4f(3)g(2) + 3f(2)^2 g(2) + 6f(2)g(3) + g(4)$$
$$\vdots$$

2.2 Lagrange Inversion Problem

Theorem 2. *(Lagrange Inversion Theorem)*

$$E_{h_0}(x) = E_f^{(-1)}(x)$$
$$h_0(n) = \sum_{\lambda \in Z(n)} (-1)^{\#\lambda} N(\lambda) \, f(\lambda)$$
$$Z(n) = P(n-1) + 1$$

Example $Z(n)$'s :
$$Z(1) = \emptyset$$
$$Z(2) = \{2\}$$
$$Z(3) = \{3, 22\}$$
$$Z(4) = \{4, 32, 222\}$$
$$Z(5) = \{5, 42, 33, 322, 2222\}$$

Example $h_0(n)$'s:
$$h_0(1) = 1$$
$$h_0(2) = -f(2)$$

$$h_0(3) = -f(3) + 3f(2)^2$$
$$h_0(4) = -f(4) + 10f(2)f(3) - 15f(2)^3$$
$$h_0(5) = -f(5) + 15f(2)f(4) + 10f(3)^2$$
$$-105f(2)^2 f(3) + 105f(2)^4$$

$$\vdots$$

2.3 Parametric Problem

Problem. Given $x = u(t), y = v(t)$. Find a general formula for

$$D_n = \frac{d^n y}{dx^n}(t)|_{t=0}.$$

Solution: Assume $u'(0) = v(0) = 1$. Note that $y = v(u^{-1}(x))$. Thus, if $u(t) = E_f(t)$ and $v(t) = E_g(t)$ then

$$y = v(u^{-1}(x)) = E_g(E_{f^{-1}}(x)). \tag{11}$$

Therefore if $E_h(x) = E_g(E_{f^{-1}})(x)$ then $h(n) = v_n(t)$.

Theorem 3. *Parametric Theorem*

$$E_h(x) = E_g(E_f^{(-1)}(x))$$

$$h(n) = \sum_{\lambda \in V(n)} (-1)^{\#\lambda - r} N(\lambda)\ f(\lambda)g(r+1)$$

$$V(n) = \left\{ \lambda : \sum_{j=1}^{\#\lambda} \max(\lambda_j - 1, 1) = n - 1 \right\}$$

Example $V(n)$'s: $V(1) = \emptyset$
$V(2) = \{2, 1\}$
$V(3) = \{3, 22, 21, 11\}$
$V(4) = \{4, 32, 31, 222, 221, 211, 111\}$

$$\vdots$$

Example $h(n)$'s: $h(0) = 1$
$h(1) = g(1)$
$h(2) = -f(2)g(1) + g(2)$
$h(3) = -f(3)g(1) + 3f(2)^2 g(1) - 3f(2)g(2) + g(3)$
$h(4) = -f(4)g(1) + 10f(2)f(3)g(1) - 4f(3)g(2)$
$$-15f(2)^3 g(1) + 15f(2)^2 g(2) - 6f(2)g(3) + g(4)$$

$$\vdots$$

3   Example: Spectral Shape of the (j, k) Ensembles

The spectral shape of the (j, k) ensemble is defined parametrically in Equations 1-5. The degree of tangency can be found by taking derivatives of the spectral shape and of the entropy function. At $\theta = \frac{1}{2}$, derivatives of the entropy function are given by Equation 12.

$$H^{(i)}(1/2) = \begin{cases} 0 & i \text{ odd} \\ 2^i(i-2)! & i \text{ even} \end{cases} \tag{12}$$

Using Theorem 3, we can calculate the derivatives of the spectral shape for the (j, k) ensemble at the point $\theta = \frac{1}{2}$ However, these derivatives grow exponentially, much like the derivatives of the entropy function. Rather than tabulate the derivatives, then, we will tabulate the ratio of the nth derivative of the spectral shape of the ensemble to the nth derivative of the entropy function. This is shown in Table 2. Note that each element is an integer, and if it is decreased by one, is a multiple of $j(k-1)$. In particular, the kth derivative is always $(-j(k-1)+1)H^{(k)}(1/2)$.

Table 2. Ratios of the Derivatives of the (j, k) Ensemble to those of the Entropy Functions

Derivative	Ensemble						
n	(2,4)	(3,6)	(4,8)	(5,10)	(6,12)	(7,14)	(8,16)
2	1	1	1	1	1	1	1
4	-5	1	1	1	1	1	1
6	31	-14	1	1	1	1	1
8	-209	1	-27	1	1	1	1
10	1471	136	1	-44	1	1	1
12	-10625	-164	1	1	-65	1	1
14	78079	-1364	365	1	1	-90	1
16	-580865	3361	-419	1	1	1	-119
18	4361215	12496	1	766	1	1	1
20	-32978945	-53864	-5319	-854	1	1	1

4   Example: Spectral Shape of the $(*, k)$ Ensembles

The spectral shape of the $(*, k)$ is given in Equation 6. We seek the nth derivative of $E(\theta)$ with respect to θ at $\theta = \frac{1}{2}$.
     Let

$$y(s) = \log(s) \tag{13}$$

$$x(\theta) = \frac{1 + (1 - 2\theta)^k}{2} \tag{14}$$

The derivatives of $E(\theta) = H(\theta) + y(x(\theta))$ can be found using Theorem 1.

The complete formula gets very complicated, but luckily we are only interested in what happens at $\theta = \frac{1}{2}$. With $x\left(\frac{1}{2}\right) = \frac{1}{2}$, it is simple to show that

$$y_i\left(\frac{1}{2}\right) = y_i = (-2)^i(i-1)! \tag{15}$$

$$x_i\left(\frac{1}{2}\right) = x_i = \begin{cases} 0 & i \neq k \\ -(-2)^{k-1}k! & i = k \end{cases} \tag{16}$$

Therefore, if n is not a multiple of k, the nth derivative of $E(\theta)$ at $\theta = \frac{1}{2}$ will be precisely $H^{(n)}\left(\frac{1}{2}\right)$, as there will be no non-zero terms in the summation.
If there is an integer i such that $ki = n$,

$$\frac{d^n E}{d\theta^n}\Big|_{\theta=\frac{1}{2}} = H^{(n)}\left(\frac{1}{2}\right) + (1-R)(-1)^{i+n}2^n(n-1)!k \tag{17}$$

If we combine this with Equation 12, we find that

$$\frac{d^n E}{d\theta^n}\Big|_{\theta=\frac{1}{2}} = \begin{cases} H^{(n)}\left(\frac{1}{2}\right) & n \text{ not a multiple of } k \\ H^{(n)}\left(\frac{1}{2}\right)\left(1+(-1)^{n/k}(1-R)k(n-1)\right) & n \text{ even and a multiple of } k \\ (1-R)(-1)^{n/k+n}2^n(n-1)!k & n \text{ odd and a multiple of } k \end{cases} \tag{18}$$

For rate 1/2 codes, the ratios between the derivatives of the spectral shape and those of the entropy function are given in table 3.

Table 3. Ratios of the Derivatives of the $(*, k)$ Ensemble to those of the Entropy Functions

Derivative	k							
n	2	4	6	8	10	12	14	16
2	0	1	1	1	1	1	1	1
4	4	-5	1	1	1	1	1	1
6	-4	1	-14	1	1	1	1	1
8	8	15	1	-27	1	1	1	1
10	-8	1	1	1	-44	1	1	1
12	12	-21	34	1	1	-65	1	1
14	-12	1	1	1	1	1	-90	1
16	16	31	1	61	1	1	1	-119
18	-16	1	-50	1	1	1	1	1
20	20	-37	1	1	96	1	1	1

5   Conclusions and Future Work

Tables 2 and 3 are similar in many ways. The leading ones in each column show that the degree of tangency in each case is k. This can be explained by extension of the Pless Power Moment Identities [4]. We do not yet understand, however, why the derivatives are always whole number multiples of the entropy function.

While Faà di Bruno's formula has been studied extensively by many others [5] [6] [7] [8], we are not aware of any previous publication regarding the derivatives of functions defined parametrically. The connection between these two formulas, possibly through Lagrange's formula for function inversion, remains to be explored.

References

1. Gallager, R.G.: Low Density Parity Check Codes. MIT Press, Redmond, Washington (1963)
2. Litsyn, S.: On Ensembles of Low-Density Parity-Check Codes: Asymptotic Distance Distributions. IEEE Transactions on Information Theory (April 2002)
3. Stanley, R.P., Fomin, S.: Enumerative Combinatorics. Cambridge Studies in Advanced Mathematics, vol. 2. Cambridge University Press, Cambridge (1999)
4. Aji, S., McEliece, R., Sweatlock, S.: On the Taylor Series of Asymptotic Ensemble Weight Enumerators. In: International Symposium on Communication Theory and Applications (2007)
5. di Bruno, C.F.F.: Note sur une novelle formule du calcul differentiel. Quart. J. Mathematics 1 (1855)
6. Arbogast, L.F.A.: Du Clacul des Derivations. Levrault, Strasbourg (1800)
7. Johnson, W.P.: The Curious History of Faà di Bruno's Formula. The American Mathematical Monthly 109 (March 1855)
8. Constantine, G.M., Savits, T.H.: A Multivariate Faà Di Bruno with Applications. Transactions of the American Mathematical Society 348 (1996)

Remarks on a Sequence of Minimal Niven Numbers*

H. Fredricksen[1], E.J. Ionascu[2], F. Luca[3], and P. Stănică[1]

[1] Department of Applied Mathematics, Naval Postgraduate School,
Monterey, CA 93943, USA
{HalF,pstanica}@nps.edu
[2] Department of Mathematics, Columbus State University,
Columbus, GA 31907, USA
ionascu_eugen@colstate.edu
[3] Instituto de Matemáticas, Universidad Nacional Autónoma de México,
C.P. 58089, Morelia, Michoacán, México
fluca@matmor.unam.mx

Abstract. In this short note we introduce two new sequences defined using the sum of digits in the representation of an integer in a certain base. A connection to Niven numbers is proposed and some results are proven.

Mathematics Subject Classification: 11A07, 11B75, 11L20, 11N25, 11N37, 11Y55.

Keywords: sum of digits, sequences, Niven numbers.

1 Introduction

A positive integer n is called a *Niven number* (or a *Harshad number*) if it is divisible by the sum of its decimal digits. For instance, 2007 is a Niven number since 9 divides 2007. A *q-Niven number* is one which is divisible by the sum of its base q digits (incidentally, 2007 is also a 2-Niven number). Niven numbers have been extensively studied by various authors (see [1,2,3,4,5,7,8,10], just to cite a few of the most recent works). We let $s_q(k)$ be the sum of digits of k in base q.

In this note, we define two sequences in relation to q-Niven numbers. For a fixed but arbitrary $k \in \mathbb{N}$ and a base $q \geq 2$, we ask if there exists a q-Niven number whose sum of its digits is precisely k. Therefore it makes sense to define a_k to be the smallest positive multiple of k such that $s_q(a_k) = k$. In other words, a_k is the smallest Niven number whose sum of the digits is a given positive integer k (trivially, for every k such that $1 \leq k < q$ we have $a_k = k$). We invite the reader to check that, for instance, $a_{12} = 48$ in base 10.

---

* Work by F. L. was done in the Spring of 2007 while he visited the Naval Postgraduate School. He would like to thank this institution for its hospitality. H. F. acknowledges support from the National Security Agency under contract RMA54. Research of P. S. was supported in part by a RIP grant from Naval Postgraduate School.

S.W. Golomb et al. (Eds.): SSC 2007, LNCS 4893, pp. 162–168, 2007.
© Springer-Verlag Berlin Heidelberg 2007

In [6] we remarked that q-Niven numbers with only 0's or 1's in their q-base representation, with a fixed sum of digits, do exist. So, we define b_k as the smallest positive multiple of k, which written in base q has only 0's or 1's as digits, and in addition $s_q(b_k) = k$. Obviously, a_k and b_k depend on q, but we will not make this explicit to avoid complicating the notation. Clearly, in base 2 we have $a_k = b_k$ for all k but for $q > 2$ we actually expect a_k to be a lot smaller than b_k.

2 The Results

We start with a simple argument (which is also included in [6]) that shows that the above sequences are well defined. First we assume that k satisfies $\gcd(k, q) = 1$. By Euler's theorem, we can find t such that $q^t \equiv 1 \pmod{k}$, and then define

$$K = 1 + q^t + q^{2t} + \cdots + q^{(k-1)t}.$$

Obviously, $K \equiv 0 \pmod{k}$, and so $K = kn$ for some n and also $s_q(K) = k$. Hence, in this case, K is a Niven number whose digits in base q are only 0's and 1's and whose sum is k. This implies the existence of a_k and b_k.

If k is not coprime to q, we can assume that $k = ab$ where $\gcd(b, q) = 1$ and a divides q^n for some $n \in \mathbb{N}$. As before, we can find a multiple of b, say K, such that $s_q(K) = b$. Let $u = \max\{n, \lceil \log_q K \rceil\} + 1$, and define

$$K' = (q^u + q^{2u} + \cdots + q^{ua})K.$$

Certainly $k = ab$ is a divisor of K' and $s_q(K') = ab = k$. Therefore, a_k and b_k are well defined for every $k \in \mathbb{N}$.

However, this argument gives a large upper bound, namely of size $\exp(O(k^2))$ for a_k. In the companion paper [6], we present constructive methods by two different techniques for the binary and nonbinary cases, respectively, yielding sharp upper bounds for the numbers a_k and b_k. Here we point out a connection with the q-Niven numbers. The binary and decimal cases are the most natural cases to consider. The table below describes the sequence of minimal Niven numbers a_k for bases $q = 2, 3, 5, 7, 10$, where $k = 2, \ldots, 25$.

We remark that if m is the minimal q-Niven number corresponding to k, then $q - 1$ must divide $m - s_q(m) = kc_k - k = (c_k - 1)k$. This observation turns out to be useful in the calculation of a_k for small values of k. For instance, in base ten, a_{17} can be established easily by using this simple property: 9 has to divide $c_{17} - 1$ and so we check for c_{17} the values 10, 19, and see that 28 is the first integer of the form $9m + 1$ ($m \in \mathbb{N}$) that works.

In some cases, one can find a_n explicitly, as our next result shows. In [6] we proved the following result.

Lemma 1. *If $q > 2$, then*

$$a_{q^m} = q^m \left(2q^{\frac{q^m-1}{q-1}} - 1\right).$$

If $q = 2$, then $a_{2^m} = 2^m(2^{2^m} - 1)$.

Table 1. Values of a_k in various bases

k	base 2	base 3	base 5	base 7	base 10
2	6	4	6	8	110
3	21	15	27	9	12
4	60	8	8	16	112
5	55	25	45	65	140
6	126	78	18	12	24
7	623	77	63	91	133
8	2040	80	24	32	152
9	1503	1449	117	27	18
10	3070	620	370	40	190
11	3839	1133	99	143	209
12	16380	2184	324	48	48
13	16367	3887	949	325	247
14	94206	4130	574	1022	266
15	96255	30615	4995	195	195
16	1048560	6560	624	832	448
17	483327	19601	2873	629	476
18	524286	177138	3114	342	198
19	1040383	58805	6099	1273	874
20	4194300	137780	15620	1700	3980
21	5767167	354291	12369	9597	399
22	165 15070	347732	12474	2398	2398
23	16252927	529253	31119	6509	1679
24	134217720	1594320	15624	2400	888
25	66584575	1417175	781225	10975	4975

The first part of the following lemma is certainly known, but we include a short proof for completeness.

Lemma 2. *Let $q \geq 2$ and k, n be positive integers. Then $s_q(nk) \leq s_q(k)s_q(n)$. In particular, $k = s_q(a_k) \leq s_q(k)\, s_q(a_k/k)$. A similar inequality holds for b_k, and both such inequalities are sharp regardless of the base q.*

Proof. Write

$$n = \sum_{i=0} n_i q^i, \qquad \text{and} \qquad k = \sum_{j=0} k_j q^j, \qquad \text{where } n_i, k_j \in \{0, 1, \ldots, q-1\},$$

for all indices i and j. Certainly, the product $nk = \sum_{i=0}\sum_{j=0} n_i k_j q^{i+j}$ is not necessarily the base q expansion of nk, as a certain value of $i + j$ may occur multiple times, or some products $n_i k_j$ may exceed q. However, $s_q(nk) \leq \sum_{i=0}\sum_{j=0} n_i k_j = s_q(k)s_q(n)$, which implies the first assertion.

Let us show that the inequalities are sharp in every base q. If $q = 2$, then letting $k = 2^m$, we get, by Lemma 1, that $a_{2^m} = 2^m(2^{2^m} - 1)$, $s_2(a_k) = 2^m$, $s_2(k) = 1$, and $s_2(a_k/k) = 2^m$, which shows that indeed $s_2(a_k) = s_2(k)s_2(a_k/k)$.

Similarly, Lemma 1 implies that this inequality is sharp for an arbitrary base q as well.   □

Let us look at the base 2 case. In [6], we have shown that

Theorem 3. *For all integers $k = 2^i - 1 \geq 3$, we have*

$$a_k \leq 2^{k+k^-} + 2^k - 2^{k-i} - 1, \tag{1}$$

where k^- is the least positive residue of $-k$ modulo i. Furthermore, the bound (1) is tight when $k = 2^i - 1$ is a Mersenne prime.

We extend the previous result in our next theorem, whose proof is similar to the proof of Theorem 3 in [6] using obvious modifications for the second claim, however we are going to include it here for the convenience of the reader. It is worth mentioning that, as a corollary of this theorem, the value of a_k is known for every k which is an even perfect number (via the characterization of the even perfect numbers due to the ancient Greeks, see Theorem 7.10 in [9]).

Theorem 4. *For all integers $k = 2^s(2^i - 1) \geq 3$, with $i, s \in \mathbb{Z}$, $i \geq 2$, $s \geq 0$, we have*

$$a_k \leq 2^s(2^{k+k^-} + 2^k - 2^{k-i} - 1), \tag{2}$$

where k^- is the least nonnegative residue of $-k$ modulo i. Furthermore, the bound (2) is tight when $2^i - 1$ is a Mersenne prime.

Proof. For the first claim, it suffices to show that the sum of binary digits of the upper bound on (2) is exactly k, and also that this number is a multiple of k.

Indeed, from the definition of k^-, we find that $k + k^- = ia$ for some positive integer a. Since

$$2^{k+k^-} + 2^k - 2^{k-i} - 1 = 2^{k-i}(2^i - 1) + 2^{ia} - 1$$
$$= (2^i - 1)(2^{k-i} + 2^{i(a-1)} + 2^{i(a-2)} + \cdots + 1),$$

we get that $2^s(2^{k+k^-} + 2^k - 2^{k-i} - 1)$ is divisible by k.

For the sum of the binary digits we have

$$s\left(2^{k+k^-} + 2^k - 2^{k-i} - 1\right) = s\left(2^{k+k^--1} + \cdots + 2 + 1 + 2^k - 2^{k-i}\right)$$
$$= s\left(2^{k+k^--1} + \cdots + 2^k + \cdots + \widehat{2^{k-i}} + \cdots + 2 + 1 + 2^k\right)$$
$$= s\left(2^{k+k^-} + 2^{k-1} + \cdots + \widehat{2^{k-i}} + \cdots + 2 + 1\right)$$
$$= k,$$

where $\hat{t}$ means that t is missing in that sum. The first claim is proved.

We now consider that $p = 2^i - 1$ is a Mersenne prime. Then we need to show that the right hand side of (2) is the smallest number that satisfies the conditions mentioned above. The divisibility condition implies that $a_k = 2^s x$ for some $x \in \mathbb{N}$. We need to show that $x = 2^{k+k^-} + 2^k - 2^{k-i} - 1$, or in other words,

x is the smallest number that has the sum of its digits in base 2 equal to k and it is divisible by p.

We know that $a_k \geq 2^k - 1$. Let us denote by m the first positive integer with the property that

$$2^{k+m} - 1 \equiv 2^{j_1} + 2^{j_2} + ... + 2^{j_m} \pmod{p} \tag{3}$$

for some $0 \leq j_1 < ... < j_m \leq k+m-2$. Notice that any other $m' > m$ will have this property and if we denote by $y = 2^{j_1} + 2^{j_2} + ... + 2^{j_m}$ the $a_k = 2^s(2^{k+m} - 1 - y)$ where $j_1, j_2, ..., j_m$ are chosen to maximize y. Because

$$x = 2^{k+k^- + 1} - 1 - (2^{k-i} + 2^{k+k^- - 1} + 2^{k+k^- - 2} + ... + 2^k) \equiv 0 \pmod{p}$$

we deduce that $m \leq k^- + 1$. Let us show that $m < k^- + 1$ leads to a contradiction. It is enough to show that $m = k^-$ leads to a contradiction. $2^{k+k^-} \equiv 2^{ia} \equiv 1 \pmod{p}$. Hence $0 \equiv 2^{j_1} + 2^{j_2} + ... + 2^{j_m} \pmod{p}$. Because $2^i \equiv 1 \pmod{p}$, we can reduce all powers 2^j of 2 modulo p to powers with exponents less than or equal to $i - 1$. We get at most $m \leq i - 1$ such terms. But in this case, the sum of at least one and at most $i - 1$ distinct members of the set $\{1, 2, ..., 2^{i-1}\}$ is positive and less than the sum of all of them, which is p. So, the equality (3) is impossible in this case.

Therefore $m = k^- + 1$ and one has to choose $j_1, j_2, ..., j_m$ in order to maximize y. This means $j_m = k+m-1$, $j_{m-1} = k+m-2, ...$, and finally j_1 has to be chosen in such a way it is the greatest exponent less than k such that $2^{k+m} - 1 - y \equiv 0 \pmod{p}$. Since $j_1 = k - i$ satisfies this condition and because the multiplicative index of 2 (mod p) is i this choice is precisely the value for j_1 which maximizes y. □

Next, we find by elementary methods an upper bound on a_k.

Theorem 5. *If k is a 2-Niven number, then*

$$a_k \leq k \frac{2^{i_s(k/s+1)} - 1}{2^{i_s} - 1},$$

where $s = s_2(k)$ and i_s is the largest nonzero binary digit of k. Moreover, the equality $s_2(a_k) = s_2(k)s_2(a_k/k)$ holds for at least

$$2\log 2 \frac{N}{\log N} + O\left(\frac{N}{(\log N)^{9/8}}\right)$$

integers $k \leq N$.

Proof. The observation allowing us to construct a multiple kd_k of k such that $s_2(kd_k) = k$ out of any 2-Niven number k, is to observe that we may choose d_k such that if $s_2(kd_k) = k$, then $s_2(d_k) = k/s_2(k)$. Thus, if

$$k = \sum_{i=0}^{N} k_i 2^i \quad \text{and} \quad d_k = \sum_{j=0}^{K} n_j 2^j,$$

then $k\,d_k = \sum_{i=0}^{N} \sum_{j=0}^{K} k_i n_j 2^{i+j}$. The equality holds if this is indeed the binary expansion of $k\,d_k$, that is, if $i+j$ are all distinct for all choices of i and j such that $k_i n_j \neq 0$.

This argument gives us a way to generate d_k. Let $k_{i_1}, k_{i_2}, \ldots, k_{i_s}$ be all the nonzero binary digits of k, where $s = s_2(k)$. Put $m = k/s$. Recall that d_k must be odd, and so the least nonzero digit of d_k is 1. We shall define a sequence of disjoint sets in the following way. Set $d_1 = 0$, and

$$A_1 = \{i_1, i_2, \ldots, i_s\}.$$

Now, let $d_2 = \min\{d \in \mathbb{N} \mid d - i_1 + i_\ell \notin A_1, \ell = 1, \ldots, s\} - i_1$ and set

$$A_2 = \{d_2 + i_1, d_2 + i_2, \ldots, d_2 + i_s\}.$$

Next, let $d_3 = \min\{d \in \mathbb{N} \mid d - i_1 + i_\ell \notin A_1 \cup A_2, \ell = 1, \ldots, s\} - i_1$ and set

$$A_3 = \{d_3 + i_1, d_2 + i_2, \ldots, d_3 + i_s\}.$$

Continue the process until we reach $d_m = \min\{d \in \mathbb{N} \mid d - i_1 + i_\ell \notin A_1 \cup A_2 \cup \cdots \cup A_{m-1}, \ell = 1, \ldots, s\} - i_1$ and set

$$A_m = \{d_m + i_1, d_m + i_2, \ldots, d_m + i_s\}.$$

Further, we define

$$d_k = 2^{d_1} + 2^{d_2} + 2^{d_3} + \cdots + 2^{d_m}. \tag{4}$$

Next, observe that

$$k\,d_k = \sum_{\ell=1}^{s} 2^{i_l} \sum_{p=1}^{m} 2^{d_p} = \sum_{r=1}^{m} \sum_{t \in A_r} 2^t,$$

and so the binary sum of digits of $k\,d_k$ is $s_2(k\,d_k) \leq \sum_{r=1}^{m} \sum_{t \in A_r} 1 = ms = k$, since the cardinality of each partition set A_r is s.

Regarding the bound on a_k, the worst case that can arise would be to take $d_j = j i_s$ at every step in the construction of the sequence of sets A_j. Thus, an upper bound for a_k is given by

$$a_k \leq 1 + 2^{i_s} + \cdots + 2^{m \cdot i_s} = \frac{2^{i_s(m+1)} - 1}{2^{i_s} - 1}.$$

We now observe that if the equality $s_2(a_k) = s_2(k) s_2(a_k/k)$ holds, since $k = s_2(a_k)$, then k is a 2-Niven number. Finally, the last estimate follows from the previous observation together with Theorem D of [7] concerning the counting function of the 2-Niven numbers. $\qquad \square$

Let us consider an example to illustrate the approach of Theorem 5. Let $n = 34 = 2^1 + 2^5$. Thus, $s = 2$, $i_1 = 1, i_2 = 5$. Now, the sequence of the sets A_i, where $i = 1, \ldots, \frac{34}{2} = 17$ runs as follows:

$$\{1,5\}, \{2,6\}, \{3,7\}, \{4,8\}, \{9,13\}, \{10,14\}, \{11,15\}, \{12,16\}, \{17,21\},$$
$$\{18,22\}, \{19,23\}, \{20,24\}, \{25,29\}, \{26,30\}, \{27,31\}, \{28,32\}, \{33,37\}.$$

Subtracting $i_1 = 1$ from the smallest element of each set A_i, we can define

$$d_{34} = 2^0 + 2^1 + 2^2 + 2^3 + 2^8 + 2^9 + 2^{10} + 2^{11} + 2^{16} + 2^{17} + 2^{18} + 2^{19} + 2^{24} + 2^{25} + 2^{26} + 2^{27} + 2^{32}.$$

It is immediate that $s_2(34\,d_{34}) = 34$ (we invite the reader to check that a_{34} is strictly smaller than d_{34}).

One can introduce a new restriction on Niven numbers in the following way: we define a *strongly q-Niven number* to be a q-Niven number whose base q digits are all 0 or 1. Obviously, every 2-Niven number is a strongly 2-Niven number. Other examples include

$$q + q^2 + \cdots + q^q, \qquad \text{or} \qquad q + q^3 + q^5 + \cdots + q^{2q+1},$$

which are both strongly q-Niven numbers for any base q. The related problem of investigating the statistical properties of the strongly q-Niven numbers seems interesting and we shall pursue this elsewhere.

References

1. Cai, T.: On 2-Niven numbers and 3-Niven numbers. Fibonacci Quart. 34, 118–120 (1996)
2. Cooper, C.N., Kennedy, R.E.: On consecutive Niven numbers. Fibonacci Quart. 21, 146–151 (1993)
3. De Koninck, J.M., Doyon, N.: On the number of Niven numbers up to x. Fibonacci Quart. 41(5), 431–440 (2003)
4. De Koninck, J.M., Doyon, N., Katai, I.: On the counting function for the Niven numbers. Acta Arith. 106, 265–275 (2003)
5. Grundman, H.G.: Sequences of consecutive Niven numbers. Fibonacci Quart. 32, 174–175 (1994)
6. Fredricksen, H., Ionascu, E.J., Luca, F., Stanica, P.: Minimal Niven numbers (I), submitted (2007)
7. Mauduit, C., Pomerance, C., Sárközy, A.: On the distribution in residue classes of integers with a fixed digit sum. The Ramanujan J. 9, 45–62 (2005)
8. Mauduit, C., Sárközy, A.: On the arithmetic structure of integers whose sum of digits is fixed. Acta Arith. 81, 145–173 (1997)
9. Rosen, K.H.: Elementary Number Theory, 5th edn. (2005)
10. Vardi, I.: Niven numbers. Computational Recreations in Mathematics, 19, 28–31 (1991)

The Linear Vector Space Spanned by the Nonlinear Filter Generator

Sondre Rønjom and Tor Helleseth

The Selmer Center,
Department of Informatics, University of Bergen
PB 7800
N-5020 Bergen, Norway

Abstract. The filter generator is an important building block in many stream ciphers. The generator consists of a linear feedback shift register (LFSR) of length n and a Boolean filtering function of degree d that combines bits from the shift register and creates an output bit z_t at any time t. A new attack on stream ciphers based on linear shift registers has recently been described by the authors in [3]. This attack is modified to stream ciphers based on any linear shift register and not only for LFSRs. The focal point of this paper is to present a linear description of the filter generator in terms of matrices. The filter generator is viewed entirely in terms of powers of a unique linear operator T together with a vector representing the filtering function. It is proved that T embodies the coefficient sequences described in [3]. Thus, interesting properties of the vector space (e.g. the dimension of the equation systems) generated by the filter generator can be analysed using theory of cyclic vector spaces, which very elegantly complements analysis in terms of the roots of the LFSR.

Keywords: stream ciphers, m-sequences, cyclic vector spaces.

1 Introduction

The filter generator uses a primitive linear feedback shift register(LFSR) of length n that generates a maximal linear sequence (an m-sequence) $\{s_t\}$ of period $2^n - 1$ satisfying the recursion

$$\sum_{j=0}^{n} c_j s_{t+j} = 0, \quad c_j \in \{0,1\}$$

where $c_0 = c_n = 1$ and $g(x) \in F_2[x]$ is a primitive polynomial. The zeroes of $g(x)$ are $\beta, \beta^2, \ldots, \beta^{2^{n-1}}$ where β is a primitive element in $\mathbb{F}_{2^n}$, the finite field with 2^n elements. The non-singular matrix

$$T_1 = \begin{pmatrix} 0 & 0 & \ldots & 0 & c_0 \\ 1 & 0 & \ldots & 0 & c_1 \\ \vdots & \vdots & \ddots & \vdots & \vdots \\ 0 & 0 & \ldots & 1 & c_{n-1} \end{pmatrix},$$

S.W. Golomb et al. (Eds.): SSC 2007, LNCS 4893, pp. 169–183, 2007.

is the companion matrix of $g(x)$ and thus

$$g(T_1) = T_1^n + c_{n-1}T_1^{n-1} + c_{n-2}T_1^{n-2} + \ldots + T_1^0 = 0.$$

As $g(x)$ is irreducible and primitive, it is also the minimal polynomial for T_1, so the above matrix is clearly invertible. Let $S_0 = [s_0, s_1, \ldots, s_{n-1}]$ denote the initial state of the LFSR. Any state S_t at time t is found by taking appropriate powers of T_1 combined with the initial state

$$S_t = [s_t, s_{t+1}, \ldots, s_{t+n-1}] = [s_0, \ldots, s_{n-1}]T_1^t,$$

and the consecutive states of the LFSR are

$$S_0, S_0T_1, S_0T_1^2, \ldots, S_0T_1^t, \ldots$$

which clearly is an n-dimensional cyclic vector space.

A Boolean function in n variables is a polynomial function $f(x_0, \ldots, x_{n-1}) \in \lambda = \mathbb{F}_2[x_0, \ldots, x_{n-1}]/(x_i^2+x_i)_{0 \le i < n}$ which is a linear transformation $f : \mathbb{F}_2^n \to \mathbb{F}_2$. The algebraic degree of f is denoted by $d = \deg(f)$. At each time t, a keystream bit $z_t \in \mathbb{F}_2$ is calculated as a function of some bits of the LFSR state S_t

$$\begin{aligned} z_t &= f(s_t, \ldots, s_{t+n-1}) \\ &= f([s_0, \ldots, s_{n-1}]T_1^t) \\ &= f_t(s_0, \ldots, s_{n-1}), \end{aligned}$$

where $f_t(s_0, \ldots, s_{n-1})$ is a nonlinear function of degree d relating the initial state S_0 to the keystream bit z_t.

In a recent paper [3] Rønjom and Helleseth present a new attack that reconstructs the initial state $(s_0, s_1, \ldots, s_{n-1})$ of the binary filter generator using D keystream bits with complexity $O(D)$, where $D = \sum_{i=1}^{d} \binom{n}{i}$, after a precomputation of complexity $O(D(log_2 D)^3)$. If L is the linear complexity of the keystream then sometimes D can be replaced by L in these complexity estimates.

The main idea behind their attack is to select a polynomial that generates z_t, say the polynomial $p(x)$ of degree $D = \sum_{i=1}^{d} \binom{n}{i}$ consisting of all zeroes β^J where Hamming weight $1 \le wt(J) \le d = \deg(f)$, and an irreducible polynomial $k(x)$ of degree n dividing $p(x)$, such that $p^*(x) = p(x)/k(x) = \sum_{j=0}^{D-n} p_j x^j$ does not generate z_t.

In [3] the polynomial $k(x)$ was taken to be $g(x)$ and computing

$$\sum_{j=0}^{D-n} p_j z_{t+j} = \sum_{j=0}^{D-n} p_j f_{t+j}(s_0, s_1, \ldots, s_{n-1}) \tag{1}$$

for $t = 0, 1, \ldots, n - 1$ gives a nonsingular $n \times n$ equation system in the initial state bits $s_0, s_1, \ldots, s_{n-1}$ (as long as $p(x)/k(x)$ does not generate z_t, in which (unlikely) case one has to select a different polynomial for $k(x)$ and do some modifications).

The reason why this works can be easily explained in the following way. For any $S = \{s_0, s_1, \ldots, s_{r-1}\} \subset \{0, 1, \ldots, n-1\}$ let $s_I = s_{i_0} s_{i_1} \cdots s_{i_{r-1}}$. Let $K_{I,t}$ be the coefficient for the monomial s_I in the corresponding equation at time t. Then the system of equations can be represented in terms of the coefficient sequences $K_{I,t}$ as

$$z_t = \sum_I s_I K_{I,t}. \tag{2}$$

The method in [3] shows that all coefficient sequences $K_{I,t}$ where $|I| \geq 2$ corresponding to all nonlinear terms obey the same linear recursion with characteristic polynomial $p^*(x) = \sum_{j=0}^{D-n} p_j x^j$ with zeros β^J where $2 \leq wt(J) \leq d = \deg(f)$. Thus the right hand side in (1) is linear, leading to an $n \times n$ linear equation system in the initial state.

One should observe that the underlying idea behind the filter generator attack is that any keystream z_t coming from a filter generator can be written as

$$z_t = \sum_{i=0}^{2^n-1} c_i \beta^{it}.$$

In the case above (with $k(x) = g(x)$), then

$$\sum_{j=0}^{D-n} p_j z_{t+j} = \sum_{\{i \mid g(\beta^i) = 0\}} c_i \beta^i = Tr(c\beta^i)$$

a linear combination of the bits in the initial state. Since the left hand side is known it is straightforward to compute the initial state. This is done detailed and rather explicit in [2].

In [4] the attack is extended to filter generators over $GF(2^m)$. For an introduction to results on algebraic attacks the reader is referred to [1], [6], and [3] for a comparison of the attack described in this paper with fast algebraic attacks.

The main purpose in this paper is to give another view of the attack in [3] by studying a matrix T that describes the filter generator in a natural way. We show how the elements in the matrix T are related to the coefficient sequences.

In the next section we explain how to extend the attack in [3] on a filter generator based on a linear feedback shift register to a filter generator based on a general linear transformation. In Section 3 we introduce a matrix T related to the filter generator and show some of its basic properties. In Section 4 we study the matrix T and its role in the attack on the filter generator.

2   Attacking the LSM Filter Generator

In this section we modify our attack to include any linear shift register, or any linear state machine(LSM), and not only linear *feedback* shift registers. The main idea is that the attack on any LSM can in a natural way be reduced to the attack of a filter generator with a linear feedback shift register.

Let S_t be the n-bit state of the linear shift register at time t. Let A be the transformation matrix of the linear shift register, i.e., $S_{t+1} = S_t A$. We assume that the matrix A is nonsingular since otherwise the shift register is considered to be degenerated. The minimal polynomial of an $n \times n$ matrix A is the nonzero monic polynomial $m_A(x)$ of smallest degree such that $m_A(A) = 0$. The characteristic polynomial $g(x) = \sum_{i=0}^{n} g_i x^i$ of A is defined by

$$g(x) = \det(A + xI).$$

It is well known that $g(A) = 0$. We will assume that A is a non-derogatory matrix (i.e., a matrix for which its minimal polynomial $m(x)$ equals its characteristic polynomial).

Then if S_0 is the initial state of the LSM we have

$$S_t = S_0 A^t.$$

Let $f(x_0, x_1, \cdots, x_{n-1})$ be the Boolean function of degree d implementing the filter generator. Then the output bit at time t is given by

$$f(S_t) = z_t.$$

It is well known that for any non-derogatory matrix A it holds that

$$A = MT_1 M^{-1}$$

where M is a nonsingular $n \times n$ matrix and T_1 is the companion matrix of $g(x)$. Note that we have

$$\begin{aligned}
S_t &= S_0 A^t \\
&= S_0 (MT_1 M^{-1})^t \\
&= S_0 MT_1^t M^{-1}
\end{aligned}$$

which implies

$$S_t M = (S_0 M) T_1^t.$$

Let $U_t = S_t M$ (and therefore $S_t = U_t M^{-1}$) and observe that this leads to

$$U_t = U_0 T_1^t.$$

We define and compute the Boolean function $\hat{f}$ of degree d by

$$\hat{f}(x_0, x_1, \cdots, x_{n-1}) = f((x_0, x_1, \cdots, x_{n-1}) M^{-1}).$$

Then this leads to

$$\begin{aligned}
z_t &= f(S_t) \\
&= f(U_t M^{-1}) \\
&= \hat{f}(U_t).
\end{aligned}$$

Let $U_0 = (u_0, u_1, \cdots, u_{n-1})$ and observe that U_t is generated by a linear feedback shift register with companion matrix T_1 and $U_t = (u_t, u_{t+1}, \cdots, u_{t+n-1})$, where the sequence $\{u_t\}$ has characteristic polynomial $g(x)$. The corresponding output bit at time t is $z_t = \hat{f}(U_t)$.

We can therefore apply the attack for the filter generator based on a linear feedback shift register with generator polynomial $g(x)$ and filter function $\hat{f}$. We write

$$z_t = \hat{f}(U_t) = \sum_I K_{I,t} u_I.$$

The attack works as before by calculating the polynomial $p^*(x) = \sum_{j=0}^{D-n} p_j x^j$ that generates all coefficient sequences of degree at least 2. Then

$$\sum_{j=0}^{D-n} p_j z_{t+j} = \sum_{j=0}^{D-n} p_j \hat{f}_{t+j} = \hat{f}_t^*$$

for $t = 0, 1, \cdots, n-1$.

This provides a linear $n \times n$ equation system in $u_0, u_1, \cdots, u_{n-1}$. Therefore, we solve this system and determine $u_0, u_1, \cdots, u_{n-1}$. Thereafter, we determine the initial state $S_0 = (s_0, s_1, \cdots, s_{n-1})$ of the linear shift register by

$$S_0 = U_0 M^{-1}.$$

The extra cost involved is the computation of $\hat{f}$ which depends on the Boolean function. The additional matrix multiplication cost in the final step does not change the overall complexity, as n is small, say typically $n = 128$, in practice.

3   T-matrix

In this section we consider a matrix representation of the filter generator. From the previous section we may assume that the sequence generator is a linear feedback shift register. The purpose of this section is to describe a nonsingular $(2^n - 1) \times (2^n - 1)$ matrix T which is unique for each register and invariant of the filter function. The matrix will often be truncated to a $D \times D$ matrix (also called T). The columns of T will contain all distinct products of n linear equations up to a given degree d, and the rows will be ordered by the monomials occurring in these equations. The matrix T is then shown to be a linear operator on a vector space V with $\dim(V) = D$. Then for a filtering function $f \in \lambda$, we may span a sequence of equations $f_0, f_1, \ldots$ by right-multiplying f with powers of T, relating the keystream bits z_t to the sequence generated by the LFSR.

Let $\hat{S}_t$ denote the vector with components s_{t+I} for $I \subset \{0, 1, \ldots, n-1\}$ in some ordering, say graded reverse lexicographic. We call $\hat{S}_t$ the (extended) state of the usual n-bit state $S_t = (s_t, s_{t+1}, \ldots, s_{t+n-1})$. We illustrate the definition with an example.

Example 1. Let $g_1(x) = x^4 + x + 1 \in \mathbb{F}_2[x]$ be the generator polynomial for the LFSR. Then for $n = 4$ and $t = 0$ we have

$$\hat{S}_0 = (s_0, s_1, s_2, s_3, s_0 s_1, s_0 s_2, s_1 s_2, s_0 s_3, s_1 s_3, s_2 s_3, s_0 s_1 s_2, s_0 s_1 s_3, s_0 s_2 s_3, s_1 s_2 s_3, s_0 s_1 s_2 s_3).$$

Using the linear recursion $s_{t+4} = s_{t+1} + s_t$ or $s_4 = s_1 + s_0$ we obtain the next (extended) state by increasing all indices by one. Thus the (extended) state at time $t = 1$ is

$$\hat{S}_1 = (s_1, s_2, s_3, s_4, s_1 s_2, s_1 s_3, s_2 s_3, s_1 s_4, s_2 s_4, s_3 s_4, s_1 s_2 s_3, s_1 s_2 s_4, s_1 s_3 s_4, s_2 s_3 s_4, s_1 s_2 s_3 s_4).$$

Note that using the linear recursion of the LFSR each component in $\hat{S}_1$ is a linear combination of the components in $\hat{S}_0$. In this case we observe that the components in $\hat{S}_1$ not containing s_4 equals directly a component in $\hat{S}_0$, while the components involving s_4 can be written as

$$
\begin{aligned}
s_4 &= s_1 + s_0 \\
s_1 s_4 &= s_1 + s_0 s_1 \\
s_2 s_4 &= s_0 s_2 + s_1 s_2 \\
s_3 s_4 &= s_0 s_3 + s_1 s_3 \\
s_1 s_2 s_4 &= s_0 s_1 s_2 + s_1 s_2 \\
s_1 s_3 s_4 &= s_0 s_1 s_3 + s_1 s_3 \\
s_2 s_3 s_4 &= s_0 s_2 s_3 + s_1 s_2 s_3 \\
s_1 s_2 s_3 s_4 &= s_0 s_1 s_2 s_3 + s_1 s_2 s_3.
\end{aligned}
$$

Therefore the linear transformation that takes $\hat{S}_0$ to $\hat{S}_1$ (or equivalently $\hat{S}_t$ to $\hat{S}_{t+1}$ for any integer t) can be described by the 15×15 matrix t given below. This matrix also occurred in the paper by Hawkes and Rose [6] in their study of algebraic attacks. The interesting observation to be showed later is that the elements in the powers T^t of the matrix T are equal to the coefficient sequences $K_{I,J,t}$ defined by Rønjom and Helleseth in [3] as the coefficient of s_I in $s_{t+J} = s_{t+j_0} s_{t+j_1} \cdots s_{t+j_{r-1}}$ where $J = \{j_0, j_1, \ldots, j_{r-1}\}$, or in other words

$$s_{t+J} = \sum_I s_I K_{I,J,t}. \tag{3}$$

Thus the transformation matrix T given by $\hat{S}_{t+1} = \hat{S}_t T$ has more consequences for attacking the filter generator than anticipated in [6]. The $(2^n - 1) \times (2^n - 1)$ matrix T is formed by taking all distinct products of the s_i's of degree $1 \leq i \leq n$ as columns and represent these as products of the s_i's of degree $0 \leq i \leq n - 1$. The columns are arranged as the rows but all indices are increased by one. The matrix is illustrated in our example for the case $n = 4$ by

$$
T = \begin{array}{c}
s_0 \\ s_1 \\ s_2 \\ s_3 \\ s_0 s_1 \\ s_0 s_2 \\ s_1 s_2 \\ s_0 s_3 \\ s_1 s_3 \\ s_2 s_3 \\ s_0 s_1 s_2 \\ s_0 s_1 s_3 \\ s_0 s_2 s_3 \\ s_1 s_2 s_3 \\ s_0 s_1 s_2 s_3
\end{array}
\left(\begin{array}{cccc|cccccc|cccc|c}
0&0&0&1&0&0&0&0&0&0&0&0&0&0&0\\
1&0&0&1&0&0&0&1&0&0&0&0&0&0&0\\
0&1&0&0&0&0&0&0&0&0&0&0&0&0&0\\
0&0&1&0&0&0&0&0&0&0&0&0&0&0&0\\
0&0&0&0&0&0&0&1&0&0&0&0&0&0&0\\
0&0&0&0&0&0&0&0&1&0&0&0&0&0&0\\
0&0&0&0&1&0&0&0&1&0&0&1&0&0&0\\
0&0&0&0&0&0&0&0&0&1&0&0&0&0&0\\
0&0&0&0&0&1&0&0&0&1&0&0&1&0&0\\
0&0&0&0&0&0&1&0&0&0&0&0&0&0&0\\
0&0&0&0&0&0&0&0&0&0&0&1&0&0&0\\
0&0&0&0&0&0&0&0&0&0&0&0&1&0&0\\
0&0&0&0&0&0&0&0&0&0&0&0&0&1&0\\
0&0&0&0&0&0&0&0&0&0&1&0&0&1&1\\
0&0&0&0&0&0&0&0&0&0&0&0&0&0&1
\end{array}\right).
$$

Notice the $\binom{n}{r} \times \binom{n}{r}, r = 1, 2, \ldots, n$ matrices along the diagonal. The matrix in the example has minimal polynomial of degree $15 = 2^n - 1$ and is in general a linear operator on the $(2^n - 1)$-dimensional vector space. Note that the columns are ordered similar to the rows but increasing all indices by 1, and in the example the ordering is therefore $s_1, s_2, s_3, s_4, s_1 s_2, s_1 s_3, s_2 s_3, s_1 s_4 \ldots$ etc.

Observe that each column in T above (say $s_1 s_4 (= s_1 + s_0 s_1)$) is written in terms of the extended previous state vector and the coefficients in front of each term represent the coefficient sequences at time $t = 1$. Furthermore, note that all columns s_I where $I \subset \{1, 2, \ldots, n-1\}$ have weight 1 since the product S_I is already the representation in terms of $\hat{S}_0$.

The motivation by Hawkes and Rose [6] for introducing the matrix T was to represent the transformation of a Boolean function $f(s_t, s_{t+1}, \ldots, s_{t+n-1}) = f_t(s_0, s_1, \ldots, s_{n-1})$ at time t to the Boolean function $f(s_{t+1}, s_{t+2}, \ldots, s_{t+n}) = f_{t+1}(s_0, s_1, \ldots, s_{n-1})$ at the next time $t + 1$.

Let

$$f(s_0, s_1, \ldots, s_{n-1}) = f_0(s_0, s_1, \ldots, s_{n-1}) = \sum_{I \subset \{0,1,\ldots,n-1\}} c_{I,f} s_I$$

where $c_{I,f} \in \{0,1\}$ depends on f. Let v_f denote the binary vector of length $2^n - 1$ with component in position I being $v_f(I) = c_{I,f}$. Then since, popular speaking, the effect of T is to increase the indices by one, this implies that the binary vector representation of $f_1(s_0, s_1, \ldots, s_{n-1}) = f_0(s_1, s_2, \ldots, s_n)$ is given by

$$v_{f_1} = T v_{f_0}.$$

Therefore, in general each output bit z_t from the filter generator leads to the equation

$$z_t = \hat{S}_0 T^t v_f. \tag{4}$$

Thus the powers of the T matrix is a handy way of describing the equation system coming from the filter generator.

Example 2. Let

$$f(s_0, s_1, s_2, s_3) = s_2 + s_0 s_1 + s_1 s_2 s_3 + s_0 s_1 s_2 s_3$$

be the filter function taking $s_t, s_{t+1}, s_{t+2}, s_{t+3}$ as input at time t and producing a keystream bit $z_t = f(s_t, s_{t+1}, s_{t+2}, s_{t+3})$. The coefficient vector of $f = f_0$ is

$$v_f = [0, 0, 1, 0, 1, 0, 0, 0, 0, 0, 0, 0, 0, 0, 1, 1].$$

Then the coefficient vector of f_1 is given by

$$v_{f_1} = T v_f = [0, 0, 0, 1, 0, 0, 1, 0, 0, 0, 0, 0, 1, 0, 1]$$

corresponding to

$$\begin{aligned}
f_1(s_0, s_1, s_2, s_3) &= s_3 + s_1 s_2 + s_2 s_3 s_4 + s_1 s_2 s_3 s_4 \\
&= s_3 + s_1 s_2 + s_0 s_2 s_3 + s_0 s_1 s_2 s_3.
\end{aligned}$$

It is interesting to note that T is also closely related to the coefficient sequences in Rønjom and Helleseth [3] and therefore can be applied to shed more light on this attack.

Let $\Omega(h(x))$ be the set of sequences generated by $h(x)$. The following lemma on the coefficient sequences in [3] is useful.

Lemma 1. *Let $g_r(x)$ be the polynomial with zeros β^J where $wt(J) = r$. Then $K_{I,J,t}$ belongs to $\Omega(g_i(x) g_{i+1}(x) \cdots g_j(x))$ where $i = |I|$ and $j = |J|$.*

In the following we relate the elements of T to certain coefficient sequences. Let I and J be subsets of $\{0, 1, \ldots, n-1\}$. We index the rows of T by subsets $I = \{i_0, i_1, \ldots, i_{r-1}\}$ corresponding to $s_I = s_{i_0} s_{i_1} \cdots s_{i_{r-1}}$. Similarly we index the columns of T by subsets $J = \{j_0, j_1, \ldots, j_{s-1}\}$ where column J represents the product s_{1+J} defined by $s_{1+j_0} s_{1+j_1} \cdots s_{1+j_{r-1}}$.

Theorem 1. *Let $T_{I,J}^t$ denote the element in row I and column J in T^t. Let $K_{I,J,t}$ be defined as the coefficient of s_I in the term s_{t+J}. Then $T_{I,J}^t = K_{I,J,t}$.*

Proof. By definition we have $\hat{S}_{t+1} = \hat{S}_t T$ and therefore that $\hat{S}_t = \hat{S}_0 T^t$. This means that column J of T^t gives a representation of s_{t+J} in terms of the components s_I in $\hat{S}_0$. It follows from (3) that

$$s_{t+J} = \sum_{I \subset \{0,1,\ldots,n-1\}} s_I K_{I,J,t}.$$

Therefore we have $T_{I,J}^t = K_{I,J,t}$. In particular $T_{I,J} = K_{I,J,1}$.   □

We next observe that the $(2^n - 1) \times (2^n - 1)$ matrix T is nonsingular.

Theorem 2. *The matrix T is nonsingular.*

Proof. It is sufficient to show that the matrices $T_1, T_2, \ldots T_n$ along the diagonal are all nonsingular. Consider T_r and suppose a linear combination of the rows of T_r is the all zero vector. Then

$$\sum_{|I|=r} a_I (T_r)_{I,J} = 0 \text{ for all subsets } J \text{ with } |J| = r .$$

In the case $J = \{j_0, j_1, \ldots, j_{r-1}\} \subset \{0, 1, \ldots, n-2\}$ then $1+J \subset \{1, 2, \ldots, n-1\}$. It follows that the column in T_r (as well as T) indexed by J (corresponding to $s_{1+j_0} s_{1+j_1} \cdots s_{1+j_{r-1}}$) contains only a single one which occurs in the row $I = 1 + J$. It follows that the rows in T_r indexed by any $I \subset \{1, 2, \ldots, n-1\}$ are linearly independent and therefore that $a_I = 0$ for these values of I.

It remains to show that $a_I = 0$ when $0 \in I$. Select a row in T_r indexed by I such that $|I| = r$ and $0 \in I$. Suppose a column indexed by J (corresponding to $s_{1+j_0} s_{1+j_1} \cdots s_{1+j_{r-1}}$) contains the term s_I. This requires that $n - 1 \in J$ and thus this column can be written $s_{1+j_0} s_{1+j_1} \cdots s_{1+j_{r-2}} s_n$. Replacing s_n by the linear combination of $s_0, s_1, \ldots, s_{n-1}$ corresponding to the LFSR gives only one term containing s_0, namely the term $s_0 s_{1+j_0} s_{1+j_1} \cdots s_{1+j_{r-2}}$. Since no other column with a different J can provide this term, it follows that each row in T_r indexed by an I containing 0 has a single one in this row. Hence, these rows are linearly independent and we obtain that $a_I = 0$ also for any I containing 0. We therefore obtain that T_r has full rank $\binom{n}{r}$ for any $r = 1, 2, \ldots, n$ and we conclude that T has full rank. □

Theorem 3. *The minimal polynomial $m_{T_r}(x)$ and characteristic polynomial $c_{T_r}(x)$ of the square $\binom{n}{r} \times \binom{n}{r}$ matrix T_r are equal. Moreover, we have that*

$$m_{T_r}(x) = g_r(x) = \prod_{I, wt(I)=r} (x + \beta^I). \tag{5}$$

Consequently, we have that

$$m_T(x) = c_T(x) = g_1(x) g_2(x) \cdots g_n(x) = x^{2^n - 1} + 1.$$

Proof. We will show that the $g_r(x)$ with its zeros being α^j, where $wt(j) = r$, is the characteristic polynomial of the matrix T_r satisfying $T_{I,J}^t = K_{I,J,t}$. The degree of $g_r(x)$ is therefore $\binom{n}{r}$. Since $g_r(T_r) = 0$, it is sufficient to show that $c(T_r) = 0$ is impossible for any polynomial $c(x)$ of degree less than $\binom{n}{r}$. So we assume that $c(T_r) = 0$ for a polynomial $c(x)$ of degree less than $\binom{n}{r}$ and show that this leads to a contradiction.

Let $c(x) = \sum_{l=0}^{d'} c_l x^l$ such that

$$\sum_{l=0}^{d'} c_l T_r^l = 0.$$

Multiplying by T_r^t implies, since $T_{I,J,t} = K_{I,J,t}$ for $t = 0, 1, \ldots$ and all I, J with $|I| = |J| = r$, that

$$\sum_{l=0}^{d'} c_l K_{I,J,t+l} = 0.$$

For $J = \{j_0, j_1, \cdots, j_{r-1}\}$, let $s_{t+J} = s_{t+j_0} s_{t+j_1} \cdots s_{t+j_{r-1}}$. Then by definition,

$$s_{t+J} = \sum_I K_{I,J,t} s_I = 0.$$

We have,

$$\sum_{l=0}^{d'} c_l s_{t+l+J} = \sum_{l=0}^{d'} c_l \sum_I K_{I,J,t+l} s_I \tag{6}$$

$$= \sum_I (\sum_{l=0}^{d'} c_l K_{I,J,t+l}) s_I$$

$$= 0.$$

To simplify the proof we select a special J by $J = \{0, 1, \ldots, r-1\}$. Expanding $s_t = Tr(\alpha^t)$, we can write

$$s_{t+J} = \sum_{L=0}^{2^n-2} A_L \alpha^{tL}.$$

The coefficient A_L, where $L = 2^{l_0} + 2^{l_1} + \cdots + 2^{l_{r-1}}$ has weight r, is known by Rueppel [7] to be:

$$A_L = \det \begin{bmatrix} 1 & 1 & \cdots & 1 \\ \alpha_0 & \alpha_1 & \cdots & \alpha_{r-1} \\ \alpha_0^2 & \alpha_1^2 & \cdots & \alpha_{r-1}^2 \\ \vdots & \vdots & & \vdots \\ \alpha_0^{r-1} & \alpha_1^{r-1} & \cdots & \alpha_{r-1}^{r-1} \end{bmatrix} = \prod_{0 \le i < j < r} (\alpha_j + \alpha_i)$$

where $\alpha_i = \alpha^{2^{l_i}}$ for $i = 0, 1 \ldots, r-1$. Since the l_i's are distinct, we obtain $A_L \neq 0$.

In particular, we have

$$s_{t+l+J} = \sum_{L=0}^{2^n-2} A_L \alpha^{(t+l)L}$$

$$= \sum_{L=0}^{2^n-2} \alpha^{Ll} A_L \alpha^{tL}.$$

This leads to,

$$\sum_{l=0}^{d'} c_l s_{t+l+J} = \sum_{l=0}^{d'} c_l \sum_{L=0}^{2^n-2} \alpha^{Ll} A_L \alpha^{tL}$$

$$= \sum_{L=0}^{2^n-2} c(\alpha^L) A_L \alpha^{tL}$$

From (6) it follows that all coefficients $c(\alpha^L) A_L$ have to be zero. Since $A_L \neq 0$ for all integers of Hamming weight r, it means that $c(\alpha^L) = 0$ for these $\binom{n}{r}$ values of L. This contradicts that the degree of $c(x)$ is less than $\binom{n}{r}$, so we arrive at the conclusion that $g_r(x)$ must be the minimal polynomial of T_r.   □

4   The T-matrix and Attacking the Filter Generator

For positive numbers d and $D = \sum_{i=1}^{d} \binom{n}{i}$ the $D \times D$ (truncated) matrix T has the following block diagonal form

$$T = \begin{pmatrix} T_1 & E_{1,2} & E_{1,3} & \cdots & & E_{1,d} \\ 0 & T_2 & E_{2,3} & \cdots & & E_{2,d} \\ 0 & 0 & T_3 & \cdots & & E_{3,d} \\ \vdots & \vdots & \vdots & \ddots & \vdots & \vdots \\ 0 & 0 & 0 & \cdots & T_{d-1} & E_{d-1,d} \\ 0 & 0 & 0 & \cdots & 0 & T_d \end{pmatrix}$$

where the T_i's are square $\binom{n}{i} \times \binom{n}{i}$ matrices and the $E_{i,j}$'s are $\binom{n}{i} \times \binom{n}{j}$ matrices. Note that for a Boolean function of degree d we may without loss of generality work with a truncated T.

For any polynomial $h(x) = \sum_j h_j x^j$ the corresponding matrix $h(T)$ has a value in row I and column J given by

$$h(T)_{I,J} = \sum_j h_j K_{I,J,t}.$$

In particular if $g_i(x)|h(x)$ then $h(T)_{I,J} = 0$ for any I with $i = |I|$. In particular if we let $p_2(x) = g_2(x)g_3(x)\cdots g_d(x)$ then $p_2(T)$ is zero except in the first n rows. Therefore, the resulting matrix $T' = p_2(T)$ is

$$p_2(T) = T' = \begin{pmatrix} p_2(T_1) & E'_{1,2} & E'_{1,3} & \cdots & & E'_{1,d} \\ 0 & 0 & 0 & \cdots & & 0 \\ 0 & 0 & 0 & \cdots & & 0 \\ \vdots & \vdots & \vdots & \ddots & \vdots & \vdots \\ 0 & 0 & 0 & \cdots & 0 & 0 \\ 0 & 0 & 0 & \cdots & 0 & 0 \end{pmatrix}$$

in which only the $n \times D$ upper part is nonzero. Clearly $rank(p_2(T)) = n$ since $m_{p_2(T)}(x) = m_T(x)/p_2(x) = g_1(x)$. The properties of the matrix T should be compared with the representation described in [3], as they are the same thing viewed differently.

Example 3. The minimal polynomial $m_T(x) = g_1(x)g_2(x)g_3(x)g_4(x)$ of T in Example 1 is $x^{15} + 1$. Since $g_1(x) = x^4 + x + 1$, we compute $p_2(x) = m_T(x)/(x^4 + x + 1) = (x^4 + x^3 + x^2 + x + 1)(x^2 + x + 1)(x^4 + x^3 + 1)(x + 1) = x^{11} + x^7 + x^8 + x^5 + x^3 + x^2 + x + 1)$. Hence, $p_2(T) = I + T + T^2 + T^3 + T^5 + T^7 + T^8 + T^{11} = T'$ which is the matrix

$$T' = p_2(T) = \begin{array}{c} s_0 \\ s_1 \\ s_2 \\ s_3 \end{array} \begin{pmatrix} 1\,0\,0\,0\,1\,0\,0\,1\,0\,1\,1\,1\,0\,1\,0 \\ 0\,0\,1\,1\,0\,1\,0\,0\,0\,1\,0\,0\,1\,1\,0 \\ 1\,1\,1\,0\,0\,0\,1\,0\,1\,1\,1\,0\,1\,0\,0 \\ 0\,1\,1\,1\,0\,0\,0\,0\,1\,0\,0\,1\,1\,0 \end{pmatrix}$$

where we have removed the last 11 rows which are zero in order to save space.

Let v_f denote the length D support vector for a function $f(s_0, \ldots, s_{n-1})$ of degree d where the coefficients are ordered in the same order as the columns of T, and therefore in the same order as the expanded LFSR state S_t satisfying

$$\hat{S}_t T = \hat{S}_{t+1}, \hat{S}_t T^2 = \hat{S}_{t+2}, \ldots, \hat{S}_t T^t = \hat{S}_{2t}, \ldots$$

Since a keystream bit is given by $z_t = \hat{S}_t v_f$ and $z_{t+r} = \hat{S}_t T^r v_f = \hat{S}_{t+r} v_f$, a matrix relating sequence bits $s_t, \ldots s_{t+D-1}$ with keystream bits $z_t, \ldots, z_{t+D-1}$ is given by column vectors

$$A_t = \left(T^t v_f \middle| T^{t+1} v_f \middle| T^{t+2} v_f \middle| \ldots \middle| T^{t+D-1} v_f \right)$$

and thus $\hat{S}_0 A_t = \hat{S}_t A_0 = \hat{S}_0 T^t A_0 = [z_t, z_{t+1}, \ldots, z_{t+D-1}]$. The columns of the matrix A_t are the coefficient vectors of the functions studied in algebraic attacks.

Given a sequence $z_t = [z_t, z_{t+1}, \ldots, z_{t+D-1}]$ the solution $\hat{S}_t$ is found simply by computing $A_t^{-1} z_t$. However, since D is a large number, standard methods for solving the system usually exceeds the complexity of guessing the key. Let $f = f_0$, then $v_{f_1} = T v_{f_0}, v_{f_2} = T v_{f_1} = T^2 v_{f_0}, \ldots$ and so on. In [3] the authors compute linear equations $f_0^*, f_1^*, \ldots f_{n-1}^*$, where

$$f_t^* = \sum_{j=0}^{D-n} p_j f_{t+j}.$$

From (4) we have $z_t = \hat{S}_0 T^t v_f$, and we obtain for $0 \le t < n$,

$$z_t^* = \sum_{j=0}^{D-n} p_j f_{t+j} = \sum_{j=0}^{D-n} p_j z_{t+j} = \hat{S}_0 T^t p_2(T) v_f.$$

Note that $\hat{S}_t p_2(T) v_f$ is the coefficient vector of $\sum_{j=0}^{D-n} p_j K_{I,t+j} s_I$ and that $v_{f_t^*} = T^t p_2(T) v_{f_0}$. Let $p_2(T) = T'$, then

$$p_2(T) A_t = [v_{f_t^*}, T v_{f_t^*}, \ldots, T^{D-1} v_{f_t^*}],$$

is a system of D linear equations. Clearly, it suffices to compute $v' = p_2(T) v_{f_t}$ restricted to a length-n vector and then compute the columns of a $n \times n$ matrix by $v', T_1 v', \ldots, T_1^{n-1} v'$.

Example 4. Let $g_1(x)$ be as in Example 1 and let $f(x_0, x_1, x_2, x_3) = x_2 + x_0 x_1 + x_3 x_2 x_1 + x_3 x_2 x_1 x_0$ be the filter function taking $s_t, s_{t+1}, s_{t+2}, s_{t+3}$ as input at time t producing a keystream bit z_t. The coefficient vector of f is $v_f = [0, 0, 1, 0, 1, 0, 0, 0, 0, 0, 0, 0, 0, 1, 1]$. Thus $A_0 = [v_f, T v_f, T^2 v_f, \ldots, T^{14} v_f]$ is the coefficient matrix for a nonlinear system of equations of degree 4 relating the initial state bits $s_0, s_1, \ldots s_{n-1}$ with the keystream $[z_0, z_1, \ldots, z_{14}]$. The matrix A_0 is

$$A_0 = \begin{array}{c} s_0 \\ s_1 \\ s_2 \\ s_3 \\ s_1 s_0 \\ s_2 s_0 \\ s_2 s_1 \\ s_3 s_0 \\ s_3 s_1 \\ s_3 s_2 \\ s_2 s_1 s_0 \\ s_3 s_1 s_0 \\ s_3 s_2 s_0 \\ s_3 s_2 s_1 \\ s_3 s_2 s_1 s_0 \end{array} \left(\begin{array}{ccccccccccccccc} 0&0&1&0&0&1&1&1&1&0&1&0&0&0&1 \\ 0&0&1&1&1&1&0&1&0&0&0&1&0&0&1 \\ 1&0&0&1&1&1&1&0&1&0&0&0&1&0&0 \\ 0&1&0&0&1&1&1&1&0&1&0&0&0&1&0 \\ 1&0&0&0&1&0&0&1&0&1&1&0&0&0&0 \\ 0&0&0&1&1&0&1&1&0&1&1&0&0&1&0 \\ 0&1&0&0&1&0&1&1&0&0&0&0&1&0&0 \\ 0&0&0&1&0&0&1&0&1&1&0&0&0&0&1 \\ 0&0&1&1&0&1&1&0&1&1&0&0&1&0&0 \\ 0&0&1&0&0&1&0&1&1&0&0&0&0&1&0 \\ 0&0&0&1&0&1&0&0&1&1&0&1&1&1&0 \\ 0&0&1&0&1&0&0&1&1&0&1&1&1&0&0 \\ 0&1&0&1&0&0&1&1&0&1&1&1&0&0&0 \\ 1&0&1&0&0&1&1&0&1&1&1&0&0&0&0 \\ 1&1&1&1&1&1&1&1&1&1&1&1&1&1&1 \end{array} \right).$$

Thus, since we have already computed $p_2(x)$ in Example 3, we compute $A' = p_2(T) A_0$ and find a matrix

$$p_2(T) A_0 = A' = \begin{array}{c} s_0 \\ s_1 \\ s_2 \\ s_3 \end{array} \left(\begin{array}{ccccccccccccccc} 0&1&0&1&1&1&1&0&0&0&1&0&0&1&1 \\ 1&1&1&1&0&0&0&1&0&0&1&1&0&1&0 \\ 0&1&1&1&1&0&0&0&1&0&0&1&1&0&1 \\ 1&0&1&1&1&1&0&0&0&1&0&0&1&1&0 \end{array} \right)$$

where the last 11 rows have been removed in order to save space. It is easily verified that the rows of the above matrix are shifts of a sequence generated by $x^4 + x + 1$). Let $A'_{n \times n}$ denote the first n columns and rows of A' and suppose we receive a keystream sequence $z_t = (1, 0, 0, 0, 1, 0, 0, 1, 0, 0, 1, 1, 1, 1, 0)$. Following [3], we compute z_t^* for $t = 0, 1, 2, 3$ and obtain a vector $z^* = [z_0^*, z_1^*, z_2^*, z_3^*] = [1, 0, 0, 1]$. Then we find that $[s_0, s_1, s_2, s_3] = A'^{-1}_{n \times n} z^* = [1, 0, 1, 1]$.

The matrix T depends only on the generator polynomial and is thus unique for each LFSR. For simplicity we assume that the generator polynomial is primitive. Let V denote the vector space of dimension D over $\mathbb{F}_2$ and let $W \subseteq V$ denote a subspace of V. Then T is a linear transformation taking vectors $v \in W$ to $Tv \in W$. Since the characteristic polynomial $c_T(x)$ of T is known, we may decompose the vector space V completely and test the order (or minimal polynomial) of any filtering function $f \in \lambda$. Following [5], we may denote by $W = Z(v_f, T) = span\{v_f, Tv_f, T^2 v_f, \ldots\}$ the vector space spanned by v_f. Since $m_{T_i}(x) = c_{T_i}(x)$, we may write V as

$$V = V_1 \oplus V_2 \oplus \cdots \oplus V_n,$$

where any $u \in V_i$ satisfy $g_i(T)u = 0$. One may proceed by choosing any irreducible factor $g_{i,r}(x)$ of $g_i(x)$ with $d_{i,r} = \deg(g_{i,r})|n$, and identify the irreducible subspaces of V_i by

$$V_{i,r} = \{u \in V_i \mid g_{i,r}(T)u = 0\},$$

where $\dim(V_{i,r}) = d_{i,r}$. Going through all e_i irreducible factors of $g_i(x)$, we may write V_i as

$$V_i = V_{i,1} \oplus V_{i,2} \oplus \cdots \oplus V_{i,e_i},$$

and V then as

$$\begin{aligned} V &= V_1 \oplus V_2 \oplus \cdots \oplus V_n \\ &= \oplus_{i=1}^{e_1} V_{1,i} \oplus_{i=1}^{e_2} V_{2,i} \oplus_{i=1}^{e_3} \cdots \oplus_{i=1}^{e_n} V_{n,i}. \end{aligned}$$

If we select a v_f of degree d at random from V, we may write v_f as a sum of vectors from different subspaces

$$v_f = v_{f,1} + v_{f,2} + \ldots + v_{f,d},$$

where each $v_{f,i} \in V_i$ may again be decomposed into vectors from the irreducible subspaces of V_i. In the attack we compute the polynomial $p_2(x) = \prod_{i=2}^{d} g_i(x)$ and use this to compute v_f^*, which can now be written as

$$\begin{aligned} v_f^* &= p_2(T)v_f \\ &= p_2(T)(v_{f_1} + v_{f_2} + \ldots + v_{f_d} \\ &= p_2(T)v_{f_1} + p_2(T)(v_{f_2} + \ldots + v_{f_d}) \\ &= p_2(T)v_{f_1}. \end{aligned}$$

In other words, $p_2(x)$ annihilates the parts of v_f coming from $V_i, i = 1, 2, \ldots, d$, and we are left with $v_f^* = v_{f_1}^* \in V_1$. In general one may choose any factor $r(x)$ of $p(x) = \prod_{i=1}^{d} g_i(x)$ and annihilate the parts of the filtering function v_f coming from other vector spaces than the part of V annihilated by $r(x)$.

For instance, to check whether the output of a particular function contains the roots of a polynomial $t(x)$, one may compute $k(x) = m_T(x)/t(x)$ and test whether $k(T)v_f \neq 0$. Since $\ker(p_2(T)) = V \backslash V_1$ and $\dim(V \backslash V_1) = D - n$, we have that for a linear function v_f^* there will in general be $2^{\dim(V \backslash V_1)} = 2^{D-n}$ Boolean functions of degree d which satisfies $p_2(T)v = v_f^*$. It is a natural consequence since V is spanned by all linear combinations of $v_{i,j}, v_i \in V_i$, so there are 2^{D-n} combinations of $v_{2,j}, \ldots v_{d,j}$ for each 2^n vectors $v_1 \in V_1$.

5   Conclusion

We have described the attack on the filter generator presented in [3] in terms of a linear operator with entries being the coefficient sequences described in that paper. Thus, the properties of the filter generator (e.g. linear complexity) can be determined by analysing the vector space it generates, which very elegantly complements analysis in terms of the roots of the LFSR. The filter generator is ultimately linear, but by increasing the degree of both the generator polynomial and the filtering function, one may ensure that the dimension of the vector space it generates is greater than the complexity of guessing the key.

Acknowledgements

This work was supported by the Norwegian Research Council.

References

1. Canteaut, A.: Open problems related to algebraic attacks on stream ciphers. In: Ytrehus, O. (ed.) WCC 2005. LNCS, vol. 3969, pp. 120–134. Springer, Heidelberg (2006)
2. Rønjom, S., Gong, G., Helleseth, T.: On Attacks on Filtering Generators Using Linear Subspace Structure (submitted)
3. Rønjom, S., Helleseth, T.: A New Attack on the Filter Generator. IEEE Transactions on Information Theory 53(5), 1752–1758 (2007)
4. Rønjom, S., Helleseth, T.: Attacking the Filter Generator over $GF(2^m)$. In: Carlet, C., Sunar, B. (eds.) WAIFI 2007. LNCS, vol. 4547, Springer, Heidelberg (2007)
5. Kleppner, A.: The Cyclic Decomposition Theorem. Integral Equations and Operator Theory 25, 406–490 (1996)
6. Hawkes, P., Rose, G.: Rewriting variables: The complexity of fast algebraic attacks on stream ciphers. In: Franklin, M. (ed.) CRYPTO 2004. LNCS, vol. 3152, pp. 390–406. Springer, Heidelberg (2004)
7. Rueppel, R.A.: Analysis and Design of Stream Ciphers. Springer, Heidelberg (1986)

Existence of Modular Sonar Sequences of Twin-Prime Product Length

Sung-Jun Yoon and Hong-Yeop Song

School of Electrical and Electronic Engineering, Yonsei University, Seoul, Korea
{sj.yoon,hysong}@yonsei.ac.kr

Abstract. In this paper, we investigate the existence of modular sonar sequences of length v and mod v where v is a product of twin primes. For $v = 3 \cdot 5 = 15$, we have found some old and new examples by exhaustive search. However, the very next case $v = 5 \cdot 7 = 35$ is completely open, in that neither we know (have) an example, nor we prove the nonexistence. We describe simply some approach to locate a single example of modular sonar sequences of length 35 and mod 35, assuming (or hoping) that one exists.

Dedicated to Solomon W. Golomb on his 75th birthday

1 Introduction

A family of pseudorandom sequences with low cross correlation, good randomness, and large linear span has important application to code-division multiple-access (CDMA) communications and cryptology. [4] [7] [15]

It has long been conjectured that if a balanced binary sequence of period v has the ideal two-level autocorrelation, then v must be either $2^n - 1$ for some positive integer n, a prime p of type $4k+3$, or a product of twin-prime. [1] [6] [7] [8] [13] [16] It is interesting to note that those three types of integers seem to have no common property except for the above. For the lengths v other than those listed above, no such binary sequences of period v are currently known, neither the proof of non-existence of such examples are completed. [8] [13] [16] On the other hand, for each of these types of lengths, at least one easy construction for such sequences are well-known. [1] [6] [7] [8] [12] [13] [16]

In [10] and [11], Gong introduced a new design for families of binary sequences with low cross correlation, balance property, and large linear span. The key idea of this new design is to use short binary periodic sequences with two-level autocorrelation function and an interleaved structure to construct a set of long binary sequences with the desired properties. This property also has significant meaning with the application on signal detection of high-speed broad-band communication system. [10] [11]

Gong's construction [11] gives a $(v^2, v, 2v + 3)$ signal set consisting of v binary sequences of period v^2 whose out-of-phase autocorrelation and cross-correlation maximum is bounded by $2v + 3$. The construction requires two binary sequences

S.W. Golomb et al. (Eds.): SSC 2007, LNCS 4893, pp. 184–191, 2007.
© Springer-Verlag Berlin Heidelberg 2007

of period v with the ideal two-level autocorrelation, and a so-called shift sequence of $\mathbf{e} = (e_0, e_1, \cdots, e_{v-1})$ of length v defined over $\{0, 1, ..., v-1\}$. She proved that the construction works in general if there exists two sequences with the ideal autocorrelation together with a shift sequence $\mathbf{e}$ with a desired property, and specifically gave constructions for $\mathbf{e}$ in the following two cases: (1) when $v = p^n - 1$ for a prime p and a positive integer n, and (2) when $v = p$ which is a prime of type $4k + 3$. [11]

As long as binary sequences are concerned, the above construction uses two well-known types of balanced binary sequences of period v with the ideal two-level autocorrelation : (1) $v = 2^n - 1$ and (2) $v = p$ is a prime of type $4k + 3$. These two cases cover all types of two-level binary autocorrelation sequences as building blocks except for a class of two-level autocorrelation sequences of period v where v is a product $p(p+2)$ of twin primes.

We first recognized that "shift sequence" $\mathbf{e}$ in [11] is the same as modular sonar sequence of length v mod v [14]. In fact, it is essentially the same as the one given by Games in [3] for the case $v = 2^n - 1$ or $p^n - 1$, or the exponential-Welch construction in [5] [14] for the case $v = p$ of type $4k + 3$. This is in fact given in her earlier paper published in 1995. [10]

A sonar sequence $a_1, a_2, ..., a_n$ of length n over the integers $\{0, 1, ..., m-1\}$ is defined by the property that

$$a_i - a_{i+r} = a_j - a_{j+r} \quad \Longrightarrow \quad i = j, \tag{1}$$

for any i, j and r in the appropriate range. [9] When this sequence is represented as an $m \times n$ matrix (or pattern, or array) of dots and blanks, there is exactly one dot per column corresponding to the integer a_i in i-th column, and this pattern possesses the non-periodic two-dimensional ideal autocorrelation function, where the value of autocorrelation at shift (t, τ) is the number of dots matched when it is shifted horizontally by t and vertically by τ with respect to itself. [9] This property was used in the design of active sonar signals. [2] Here, a_i represents the carrier frequency at time slot i. So, m is the number of frequencies to be used in the system. In general, one would hope the sonar sequence be as long as possible given the number m of frequencies is fixed. In this sense, *known best* sonar sequences (or $m \times n$ arrays) up to $m = 100$ are listed in [14]. They started from a modular sonar sequence $\{a_i\}$ (mod m) and transform this into $\{b_i\}$ where

$$b_i = ua_i + si + c \quad (\text{mod } m) \quad 0 \le i < n, \tag{2}$$

where u is relatively prime to m, and s, c are any integers, and see if one can find a long run of empty rows to be deleted so that the resulting sonar array is *best* optimized. [14] Here, a modular sonar sequence is the same as a sonar sequence except that the condition (1) is replaced by

$$a_i - a_{i+r} = a_j - a_{j+r} \quad (\text{mod } m) \quad \Longrightarrow \quad i = j. \tag{3}$$

It is called an $m \times n$ modular sonar array, or a modular sonar sequence of length n and mod m.

To investigate the missing case, that is, the case $v = p(p+2)$, we have to find modular sonar sequences of length v mod v. Unfortunately, however, we were not able to show the non-existence, nor we could find one single example, except for the one special case $v = 15$. Note that the case $v = 15$ is very special in that it is the only case which is both a product of twin-prime and of the form $2^n - 1$. For the case $v = 15$, using any of these examples, the construction gives easily the families of $(v^2, v, 2v + 3)$ signal sets by way of the interleaved construction as in [11], since the existence of balanced binary sequence of period $v = p(p+2)$ with the ideal two-level autocorrelation is well-known [1] [7].

In the following, we will briefly describe some results for the case $v = 15$ in Section 2, and some ad-hoc tries to find an example for $v = 35$ in Section 3, all of which turned out to be not successful. Following are some open questions in this direction:

Q1: Does there exist a modular sonar sequence of length 35 and mod 35? No example is currently known and no proof of nonexistence is known either.

Q2: Find any example of modular sonar sequence of length v and mod v where $v = p(p + 2) > 15$ is a product of twin primes.

2   Case $v = 3 \times 5 = 15$

By an exhaustive search, we found all the 9000 modular sonar sequences of length 15 and mod 15. In the sense of the transformations given in (2) originally given in [14], these are classified into 5 inequivalent classes, each containing 1800 sequences. For any two sequences $\{a_i\}$ and $\{b_i\}$ in the same class, there exist some u, s, c such that $b_i = ua_i + si + c$ (mod 15) for $0 \le i < n$. There are 8 choices for u, and 15 choices for both s and c, and this counts $1800 = 8$ members of each class. Representatives of these 5 inequivalent classes are

$$\text{Class 1} : \mathbf{e}_1 = (1, 3, 1, 7, 11, 3, 14, 15, 8, 8, 13, 7, 4, 14, 2)$$
$$\text{Class 2} : \mathbf{e}_2 = (1, 1, 4, 1, 9, 7, 11, 1, 8, 2, 12, 13, 4, 6, 2)$$
$$\text{Class 3} : \mathbf{e}_3 = (1, 1, 2, 14, 2, 13, 4, 9, 13, 12, 4, 2, 11, 6, 8)$$
$$\text{Class 4} : \mathbf{e}_4 = (1, 1, 4, 9, 4, 11, 10, 8, 5, 9, 10, 1, 9, 5, 7)$$
$$\text{Class 5} : \mathbf{e}_5 = (1, 6, 12, 13, 10, 14, 7, 9, 7, 14, 10, 13, 12, 6, 1)$$

If we expand the concept of equivalence so that one is regarded to be equivalent to its mirror image (reverse reading), then Class 3 is equivalent to Class 1, and Class 4 is equivalent to Class 2. Thus, we have only 3 super-inequivalent classes: Classes 1, 2, and 5.

Figures 1 and 2 show both array forms and modular difference triangles of these three representative sequences. Here, the condition (3) can easily be checked by the fact that each row of triangle (except for the top row corresponding to the sequence itself) contains no repetitions. Here the circle denotes the ordinary difference is negative and hence converted to a positive value by adding 15. Some observations follow as Remarks.

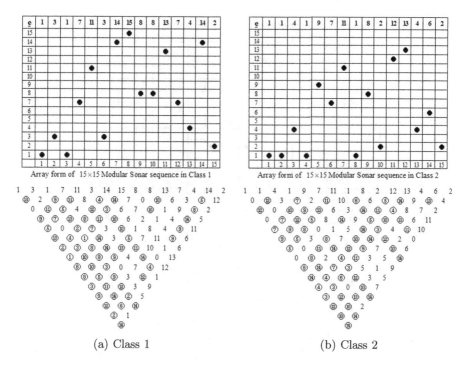

Fig. 1. Array form and modular difference triangle of sequences in Classes 1 and 2

Remark 1 (Classes 1 and 2 are NOT new). Sequences from Class 1 and Class 2 are the same as those given by Games in [3] and their transformed versions using (2). In fact, all the sequences constructed in [3] of length 15 and mod 15 are in either Class 1 or Class 2.

Remark 2 (Class 5 is new). Sequences in Class 5 are new, in the sense that no previously known algebraic constructions produce them.

Remark 3. Some sequences in Class 5 are palindromic. That is, for example, $\mathbf{e}_5 = (e_1, e_2, ..., e_{15})$ shown earlier has the property that

$$e_i = e_{14-i}, \quad 0 \le i < 15. \quad \text{[palindromic property]} \tag{4}$$

Furthermore, the first 8 terms satisfy the following:

$$|\{e_j \ominus e_{j+s} | 0 \le j < 8 - s\}| = 8 - s, \quad 1 \le s < 8, \quad \text{[modified DT property]} \tag{5}$$

where

$$
e_j \ominus e_{j+s} = \begin{cases}
15 - (e_j - e_{j+s}), & \text{if } 8 \le e_j - e_{j+s} < 15, \\
e_j - e_{j+s}, & \text{if } 0 < e_j - e_{j+s} < 8, \\
|e_j - e_{j+s}|, & \text{if } -7 \le e_j - e_{j+s} < 0, \\
15 + (e_j - e_{j+s}), & \text{if } -14 \le e_j - e_{j+s} < -7
\end{cases}
$$

and $0 < e_j \ominus e_{j+s} < 8$. This property is shown in Fig. 3.

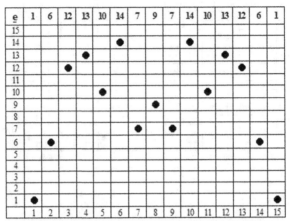

Array form of 15×15 Modular Sonar sequence in Class 5

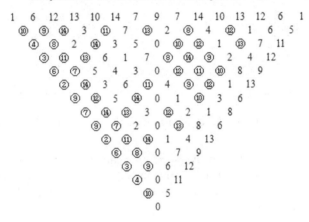

Fig. 2. Array form and modular difference triangle of sequences in Class 5

3   Case $v = 5 \times 7 = 35$

First try was to generalize Class 5 of case $v = 15$ with palindromic property and modified DT property. For the faster check in a computer search, we were able to prove the following:

Lemma 1. *Let $e = (e_0, e_1, ..., e_{v-1})$ be a sequence over $\{0, 1, ..., v-1\}$ of odd length v that is palindromic as in (4). If e is a modular sonar sequence mod v, then its first $(v+1)/2$ elements satisfy the following condition, similar to (5):*

$$|\{e_j \ominus e_{j+s} | 0 \le j < (v+1)/2 - s\}| = (v+1)/2 - s, \quad 1 \le s < (v+1)/2, \quad (6)$$

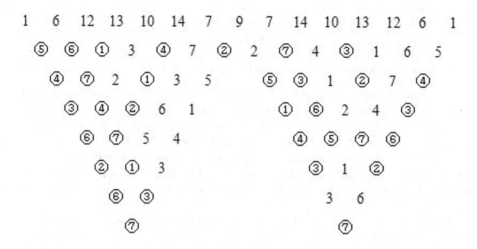

Fig. 3. Modified DT property of some sequences in Class 5

where

$$
e_j \ominus e_{j+s} = \begin{cases} v - (e_j - e_{j+s}), & \text{if} \quad (v+1)/2 \le e_j - e_{j+s} < v, \\ e_j - e_{j+s}, & \text{if} \quad 0 < e_j - e_{j+s} < (v+1)/2, \\ |e_j - e_{j+s}|, & \text{if} \quad -((v-1)/2) \le e_j - e_{j+s} < 0, \\ v + (e_j - e_{j+s}), & \text{if} \quad -(v-1) \le e_j - e_{j+s} < -((v-1)/2) \end{cases}
$$

and $0 < e_j \ominus e_{j+s} < (v+1)/2$.

Proof. If $e_j \ominus e_{j+s} = 0$, then the value $e_j - e_{j+s}$ (mod v) appears twice in row s of the original modular difference triangle, which is impossible.

Now, suppose the sequence does not satisfy the condition (6). Then, there exist $0 \le j \ne j' < (v+1)/2 - s$ such that, for some $1 \le s < (v+1)/2$, $e_{j'} \ominus e_{j'+s} = e_j \ominus e_{j+s}$. Denote

$$
a = e_j - e_{j+s}, \qquad d(s,j) = e_j \ominus e_{j+s},
$$
$$
a' = e_{j'} - e_{j'+s}, \qquad d'(s,j') = e_{j'} \ominus e_{j'+s}.
$$

From the definition of $\ominus$, then we have

$$
a \in \{v - d(s,j), d(s,j), -d(s,j), v + d(s,j)\},
$$
$$
a' \in \{v - d(s,j'), d(s,j'), -d(s,j'), v + d(s,j')\}.
$$

Since $a = v - d(s,j)$ or $a = d(s,j)$ and $a' = v - d(s,j')$ or $a' = d(s,j')$ all mod v, we have 4 cases all mod v:

$$
\begin{array}{llll}
A: & a = v - d(s,j) \text{ and } a' = v - d(s,j') & \Rightarrow & a = a', \\
B: & a = d(s,j) \text{ and } a' = d(s,j') & \Rightarrow & a = a', \\
C: & a = v - d(s,j) \text{ and } a' = d(s,j') & \Rightarrow & a = v - a', \\
D: & a = d(s,j) \text{ and } a' = v - d(s,j') & \Rightarrow & a = v - a'.
\end{array}
$$

For Cases A and B,

$$e_j - e_{j+s} = e_{j'} - e_{j'+s} \pmod{v}, \quad 0 \le j \ne j' < (v+1)/2 - s,$$

which is a contradiction. Similarly for the cases C and D. □

Hoping that the new example in Class 5 of length 15 above belongs to a general family, we have done a search for the same kind $\mathbf{e} = (e_0, e_1, ..., e_{34})$ of length 35 mod 35, by checking only the first 18 terms using Lemma 1. It took about 7 days on a PC to conclude that no such example exists.

Second try was to generalize and emulate the process of constructing the binary ideal 2-level autocorrelation sequences of period $v = p(p+2)$ from those of period p and $p+2$. [6] However, we do not have any further idea on this approach.

Third try was to generalize the method of Welch-Costas array of order $p-1$. [9] [6] There exists an integer n such that $v = p(p+2)$ is a divisor of $2^n - 1$. Consider the finite field F_{2^n} and an element β of order v in it. Successive powers of β will produce a sequence of length v over F_{2^n}. This sequence will surely satisfy the difference triangle property. However, the values are not over $\{0, 1, 2, ..., v - 1\}$ but over F_{2^n}. Therefore, we have to see if there exists a transformation that sends this sequence into that over $\{0, 1, 2, ..., v - 1\}$ preserving the difference triangle property. Various approach were tested for $n = 12$ and $v = 35$, but all failed to find any example. However, this approach finds a $2^{12} \times 35$ modular sonar array by transforming the elements of F_{2^n} into binary 12-tuples, and then by reading them (or their cyclic shifts) as binary expansions of ordinary integers. This example gives a hope for the next approach, and explicitly, it is

$$(3417, 1107, 2707, 1682, 2516, 413, 1607, 3489, 1591, 599, 3075, 2675,$$
$$2390, 3517, 468, 3268, 532, 1842, 165, 2947, 3486, 3124, 1271, 2954,$$
$$899, 199, 2151, 3684, 3352, 2647, 346, 3616, 965, 2863, 2048).$$

Fourth try was to use and modify the transformation (2) as done in [14], using the example of length 35 but mod 2^{12} found above. The goal is to find u, s, c such that the resulting transformed version has as long run of empty rows as possible to be deleted. The resulting array may not be modular, but we just tried, in vain.

References

1. Baumert, L.D.: Cyclic Difference Sets. Lecture Notes in Mathematics, vol. 182. Springer, Heidelberg (1971)
2. Costas, J.P.: Medium constraints on sonar design and performance. In: FASCON CONV. Rec., pp. 68A–68L (1975)
3. Games, R.A.: An algebraic construction of sonar sequences using M-sequences. SIAM J. Algebraic Discrete Methods 8, 753–761 (1987)
4. Golomb, S.W.: Shift Resister Sequences (revised ed.), p. 39. Aegean Park Press, Laguna Hills, CA (1982)

5. Golomb, S.W.: Algebraic constructions for Costas arrays. J. Combinatorial Theory, ser. A, 37, 13–21 (1984)
6. Golomb, S.W.: Construction of signals with favorable correlation properties. In: Keedwell, A.D. (ed.) Survey in Combinatorics. LMS Lecture Note Series, vol. 166, pp. 1–40. Cambridge University Press, Cambridge (1991)
7. Golomb, S.W., Gong, G.: Sequence Design for Good Correlation. Cambridge University Press, Cambridge (2005)
8. Golomb, S.W., Song, H.-Y.: A conjecture on the existence of cyclic Hadamard difference sets. Journal of Statistical Planning and Inference 62, 39–41 (1997)
9. Golomb, S.W., Taylor, H.: Two-dimensional synchronization patterns for minimum ambiguity. IEEE Trans. Inform. Theory IT-28, 263–272 (1982)
10. Gong, G.: Theory and applications of q-ary interleaved sequences. IEEE Trans. Inform. Theory 41(2), 400–411 (1995)
11. Gong, G.: New designs for signal sets with low cross correlation, balance property, and large linear span: GF(p) case. IEEE Trans. Inform. Theory 48, 2847–2867 (2002)
12. Hall Jr., M.: A survey of difference sets. Proc. Amer. Math. Soc. 7, 975–986 (1956)
13. Kim, J.-H.: On the Hadamard Sequences, PhD Thesis, Dept. of Electronics Engineering, Yonsei University (February 2002)
14. Moreno, O., Games, R.A., Taylor, H.: Sonar sequences from Costas arrays and the best known sonar sequences with up to 100 symbols. IEEE Trans. Inform. Theory 39, 1985–1989 (1993)
15. Simon, M.K., Omura, J.K., Scholtz, R.A., Levitt, B.K.: Spread Spectrum Communications Handbook. Computer Science Press, Rockville (1985), revised edition, McGraw-Hill (1994)
16. Song, H.-Y., Golomb, S.W.: On the existence of cyclic Hadamard difference sets. IEEE Trans. Inform. Theory 40(4), 1266–1268 (1994)

Randomness and Representation of Span n Sequences

Guang Gong

Department of Electrical and Computer Engineering
University of Waterloo
Waterloo, Ontario N2L 3G1, Canada
ggong@calliope.uwaterloo.ca

Dedicated to Solomon Golomb on his 75th Birthday

Abstract. Any span n sequences can be regarded as filtering sequences. From this observation, new randomness criteria for span n sequences are proposed. It is proved that the feedback function of a span n sequence can be represented as a composition of its trace representation, or equivalently, its discrete Fourier transform, and a permutation from the state space of the sequence to the multiplicative group of the finite field $GF(2^n)$, and vice versa. Significant enhancements for randomness of span n sequences, so that de Bruijn sequences, are illustrated by some examples.

Keywords: Nonlinear feedback shift register sequences, span n sequences, de Bruijn sequences, randomness, discrete Fourier transform.

1 Introduction

In order to seek out good pseudorandom sequence generators (PRSG) in cryptographic practice, several randomness criteria have been proposed. From Shannon's work [22], the one-time-pad is unbreakable. Hence a good PRSG should generate pseudorandom sequences with large periods to guarantee that different messages are encrypted using different key streams. In the mid 1950s, Golomb proposed the well-known three randomness postulates [12], i.e., R-1(the balance property), R-2 (the run property), and R-3 (the ideal 2-level autocorrelation). In addition, the ideal n-tuple distribution is also introduced. By the end of the 1960s, Berlekamp [2] found a decoding algorithm which can reconstruct an entire codeword from a partial known consecutive bits of the codeword. Shortly after, Massey used this algorithm in linear feedback shift register (LFSR) sequences synthesis [18]. Using the algorithm, if the length of the shortest LFSR which generates a sequence, known as the linear span or linear complexity of the sequence, is equal to n, then from any known consecutive $2n$ bits, the full period of the sequence can be reconstructed. This result imposed the large linear span criterion to PRSGs [21]. To overcome the weakness of the sole linear span criterion, two more criteria related to linear spans were introduced in the mid 1980s [21] and early 1990s [8], namely, the linear span profile and k-error linear span (or originally referred to as sphere

S.W. Golomb et al. (Eds.): SSC 2007, LNCS 4893, pp. 192–203, 2007.
© Springer-Verlag Berlin Heidelberg 2007

linear complexity) of a sequence, respectively. However, no such sequences with simple implementation as well as a smoothly increased linear span profile and/or the k-error linear span property have been found. The normalized linear span of a sequence was also introduced as a complement of the linear span measurement (see below for the definition or [15]).

With popularity of the public-key cryptography, researchers look for PRSGs whose randomness is based on computational complexity. A pseudorandom sequence is said to have *indistinguishablity* if the sequence is indistinguishable from a random bit stream generated by a truly random bit generator using any polynomial time algorithm (see [11]).

The above constitutes a rough picture of the developments of randomness criteria for PRSGs up to now. In the following, we single out one class of pseudorandom sequences, the so-called span n sequences, and summarize their known results and their potential applications in cryptology.

A binary sequence with period $N = 2^n - 1$ is called *a span n sequence* if each non-zero n-tuple $(x_0, x_1, \cdots, x_{n-1}) \in \mathbb{F}_2^n$ occurs exactly once in every period. This property was discussed in Golomb's paper [13] back in the early 1980s.

A. Pseudorandom number generators (PRNGs) and Span n Sequences: A PRNG, whenever employed in the Digital Signature Standard [16] [20] or session key generations, must generate different pseudorandom numbers at different time instances. On the other hand, each state of a span n sequence gives different binary numbers. Thus, a span n sequence can be employed as a PRNG.

B. Known Results on de Bruijn Sequences and Span n Sequences: A de Bruijn sequence of period 2^n is a binary sequence with period 2^n, which can be generated by a nonlinear feedback shift register (NLFSR) with n stages. A span n sequence can be obtained from a de Bruijn sequence of period 2^n by deleting one zero from the run of zeroes of length n, vice versa, i.e., any de Bruijn sequence can be obtained from a span n sequence by adding one zero into the zero run of length n. (A span n sequence is also referred to as a *modified de Bruijn sequence* in [17].) There are $2^{2^{n-1}-n}$ de Bruijn sequences, so does the span n sequences. (See [12].)

Chan, Games and Key [4] proved that the linear span of any de Bruijn sequence of period 2^n, denoted by L, is bounded by $2^{n-1} + n \leq L \leq 2^n - 1$ where both the lower bound and upper bound are achievable [9]. Therefore any de Bruijn sequence has a large linear span, the normalized linear span is $> 1/2$ (the normalized linear span of a sequence with period r and linear span s is defined by $\frac{s}{r}$), and satisfies the span n property. But it does not posses the 2-level autocorrelation property. On the other hand, a lower bound of the linear span of the corresponding span n sequence is dramatically dropped to n, i.e., $n \leq L \leq 2^n - 2$. No theoretical results on linear spans of span n sequences, except for m-sequences, have been established. Experimental results show that the linear span of an NLFSR span n sequence, i.e., it is not an m-sequence, varies in the range from $3n$ to $2^n - 2$ (see [17]). From a point of view in cryptographic applications, an "actual" linear span of a de Bruijn sequence should be measured in terms of the linear span of its corresponding span n sequence, because the transformation between them are deterministic. This phenomenon has been observed in 1980s. However, up to now, no sub-classes of de Bruijn sequences are discovered such that their linear spans do not differ significantly from the linear spans of their corresponding span n sequences.

The Stream Cipher Project in ECRYPT has several submissions which used NLFSRs [10]. However, we know a little about cycle structures of NLFSRs. (Most of the known results are collected in Golomb's pioneering book [12].) It has no significant progress along this line for about 5 decades.

In this paper, we discuss the cryptographic properties of span n sequences and a new approach to represent span n sequences.

2   The Basic Definitions

For a positive integer n, we denote that $N = 2^n - 1$, $q = 2^n$, $F_q = GF(q)$, and $F_2^n = \{(x_0, x_1, \cdots, x_{n-1}) \mid x_i \in F_2\}$. All the sequences that are considered in this paper are binary.

A. LFSR and NLFSR Sequences.   We say that $\{a_i\}$ is a binary sequence generated by a linear or nonlinear feedback shift register (LFSR or NLFSR) with n stages if the sequence satisfies the following recursive relation

$$a_{k+n} = f(a_k, a_{k+1}, \cdots, a_{k+n-1}), k = 0, 1, \cdots \tag{1}$$

where $(a_0, a_1, \cdots, a_{n-1})$ is an initial state, and the feedback function $f(x_0, x_1, \cdots, x_{n-1})$ is a boolean function in n variables. Any n consecutive terms of the sequence in (1), denoted by S_k,

$$S_k = (a_k, a_{k+1}, \cdots, a_{k+n-1})$$

represent a state of the shift register. If $f(x)$ is linear, say $f(x_0, x_1, \cdots, x_{n-1}) = \sum_{i=0}^{n-1} c_i x_i, c_i \in \mathbb{F}_2$, then (1) becomes $a_{k+n} = \sum_{i=0}^{n-1} c_i a_{k+i}, k = 0, 1, \cdots$. The polynomial $t(x) = x^n + \sum_{i=0}^{n-1} c_i x^i$ is called a characteristic polynomial of the sequence a. For the theory of LFSR and NLFSR, the reader is referred to [12].

Note that any boolean function f in n variables can be represented as a polynomial function from $\mathbb{F}_q$ to $\mathbb{F}_2$ in terms of the Lagrange interpolation (which has the similar formula as DFT defined below, see [15] for details). In this paper, we restrict ourselves to the case $f(0) = 0$. (For $f(0) \neq 0$, replacing $f(x)$ by $g(x) = 1 + f(x)$, then all the results obtained in this paper are applicable to the case $f(0) = 1$). Furthermore, we will not make any distinction among a boolean function and its polynomial representation in notation, whose meaning depends on the context.

B. DFT and Inverse DFT of Binary Sequences: Let $\{a_t\}$ be a binary sequence with period $N = 2^n - 1$. We also write $\{a_t\} = (a_0, a_1, \cdots, a_{N-1})$ as a vector. Let α be a primitive element in $\mathbb{F}_q (q = 2^n)$. Then the *(discrete) Fourier Transform (DFT)* of $\{a_t\}$ is defined by

$$A_k = \sum_{t=0}^{N-1} a_t \alpha^{-tk}, \ k = 0, 1, \ldots, N-1. \tag{2}$$

The inverse DFT (IDFT) is given by

$$a_t = \sum_{k=0}^{N-1} A_k \alpha^{kt}, \quad t = 0, 1, \ldots, N-1. \tag{3}$$

The sequence $\{A_k\}$ is referred to as a *spectral sequence* of $\{a_t\}$. Let $A(x) = \sum_{k=0}^{N-1} A_k x^k$. Then $a_t = A(\alpha^t)$.

Fact 1. 1. $A_{2^j k} = A_k^{2^j}, \forall j, k$.
2. $A(x)$ can be written as

$$A(x) = \sum_k Tr_1^{m_k}(A_k x^k) \tag{4}$$

where the k's are (cyclotomic) coset leaders modulo N, $m_k \mid n$ is the length of the coset which contains k, and $Tr_1^{m_k}(x)$ is a trace function from $\mathbb{F}_{2^{m_k}}$ to $\mathbb{F}_2$ (which will be written as $Tr(x)$ if $m_k = n$ in short). This is called a trace representation of $\{a_t\}$.

C. Linear Spans: The linear span of a binary sequence a is defined as the length of the shortest LFSR which generates the sequence, denoted by $LS(a)$. The corresponding characteristic polynomial is referred to as the *minimal polynomial* of the sequence. If $f(x)$ generates a sequence, then the minimal polynomial of the sequence is a divisor of $f(x)$. Any pseudorandom sequences employed in key stream generators of stream ciphers or pseudorandom number generators should have large linear spans.

For a positive integer r, an r-shift (left) of a, denoted by $L^r(a)$, is a sequence given by $a_r, a_{r+1}, \cdots$. The shift operator does not change the linear span of the resulting sequence, i.e., $LS(a) = LS(L^r(a))$. If a has period N, then the linear span of a is equal to the number of nonzeros in the spectrum of a. Furthermore, let $\{A_k'\}$ be a spectral sequence of the r-shift of a.

Fact 2. *The spectral sequences of the sequence and its r shift are related by*

$$A_k' = \alpha^{rk} A_k, \forall k.$$

D. Filtering Sequence Generators: Let $u = \{u_t\}$ be an m-sequence of period $2^n - 1$, and $0 \le k_0 < k_1 < \cdots < k_{m-1} < n$. A sequence $a = \{a_t\}$ is called a filtering sequence if

$$a_t = f(u_{k_0+t}, u_{k_1+t}, \cdots, u_{k_{m-1}+t}), t = 0, 1, \cdots \tag{5}$$

where $f(x_0, x_1, \cdots, x_{m-1})$ is a boolean function in m variables. The boolean function f is referred to as a filtering function.

3 Randomness of Span n Sequences

In this section, we first show that any span n sequence can be regarded as a filtering sequence and propose several new randomness criteria for span n

sequences, then we show that the corresponding de Bruijn sequence of a span n sequence with maximum linear span also has maximum linear span. In addition, some examples of span n sequences with maximum linear span, having very poor nonlinearity (this concept will be introduced below), are also presented in this section.

3.1   New Randomness Criteria

A known fact that has not received much attention is that a span n sequence in fact is a filtering sequence, which will be shown below. Recall that α is a primitive element in $\mathbb{F}_q$. Let $u = \{u_t\}$ be an m-sequence of period N with trace representation $Tr(x)$, i.e., $u_t = Tr(\alpha^t)$. Let $\{\alpha_0, \cdots, \alpha_{n-1}\}$ and $\{\beta_0, \cdots, \beta_{n-1}\}$ be a pair of the dual bases of $\mathbb{F}_q$ $(q = 2^n)$ over $\mathbb{F}_2$, i.e.,

$$Tr(\alpha_i \beta_j) = \begin{cases} 1 \Longleftrightarrow i = j \\ 0 \Longleftrightarrow i \neq j. \end{cases}$$

For any element $x \in \mathbb{F}_q$, we have the following relationship

$$x = \sum_{i=0}^{n-1} x_i \alpha_i, \ x_i \in \mathbb{F}_2 \Longrightarrow x_i = Tr(\beta_i x), \ i = 0, \cdots, n-1.$$

Let $a = \{a_i\}$ be a sequence with period N, and $f(x)$ be its trace representation, i.e., $a_i = f(\alpha^i), 0 \le i < N$. Let $\alpha^i = \sum_{j=0}^{n-1} c_{i,j} \alpha_j$, then $c_{i,j} = Tr(\beta_j \alpha^i)$. Let $\beta_j = \alpha^{k_j}$, then $\{Tr(\beta_j \alpha^i)\}_{i \ge 0} = L^{k_j}(u)$. Hence a can be written as

$$a_i = f(u_{k_0+i}, u_{k_1+i}, \cdots, u_{k_{n-1}+i}), i = 0, 1, \cdots. \tag{6}$$

According to (5), a is a filtering sequence where the filter function is equal to $f(x_0, x_1, \cdots, x_{n-1})$, the corresponding boolean form of $f(x)$ (here we use the same notation for $f : \mathbb{F}_q \leftarrow \mathbb{F}_2$ in both their respective polynomial form and boolean form), and the tap positions on u are given by $(k_0, k_1, \cdots, k_{n-1})$.

Since a span n sequence has period N, it can also be considered as a filtering sequence with the above form. So, when sequences of period N are employed either in stream ciphers or in pseudorandom number generators, one should consider the possibility of applying attacks on filtering generators. There exist several criteria for choosing f being a good filtering function. Thus, randomness of span n sequences with applications in cryptology should be measured not only by the randomness criteria summarized at the beginning of Section 1, but also by the following three properties which are related to filtering functions, i.e., boolean functions in n variables. Let a be a span n sequence and $f(x)$ be its trace representation.

(F1) Nonlinearity or linear resistancy of a (in terms of f): $N_a = \frac{q-c_f}{2}(= N_f)$ where $c_f = \max\{|\pm \widehat{f}(\lambda)|, \lambda \in \mathbb{F}_q\}$ where $\widehat{f}(\lambda)$ is the Hadamard transform of $f(x)$ defined by $\widehat{f}(\lambda) = \sum_{x \in \mathbb{F}_q} (-1)^{Tr(\lambda x) + f(x)}, \forall \lambda \in \mathbb{F}_q$. Then N_f should be large or equivalently, c_f, the maximum correlation between $f(x)$ and $Tr(x)$ or $f(x)$ and $Tr(x) + 1$ should be small, in order to be resistant to correlation attacks [23] and linear cryptanalysis [19].

(F2) Propagation or differential resistancy of a (in terms of f): $D_a = \frac{q - P_f}{2} (= D_f)$ where

$$P_f = \max\{|A_f(\omega)|, \omega \in \mathbb{F}_q\}$$

where $A_a(\omega)$ is the additive autocorrelation defined by

$$A_f(\omega) = \sum_{x \in \mathbb{F}_{2^n}} (-1)^{f(x) + f(x + \omega)}, \ \forall \omega \in \mathbb{F}_{2^n}.$$

The magnitude of the maximum additive autocorrelation, P_f, should be small, in order to have large differential resistance D_a, i.e., the distance between $f(x)$ and $f(x + a)$ is large (or a small change in variable x results in a big change in $f(x)$), for combatting differential cryptanalysis [3]. A link between linear cryptanalysis and differential cryptanalysis is discussed in [5].

(F3) Algebraic immunity of a (in terms of f):

$$AI_a = AI_f = \min\{deg(g) \mid g \in \mathcal{F}_n, fg = 0 \text{ or } (f + 1)g = 0\}$$

where $\mathcal{F}_n$ is the set consisting of all functions from $\mathbb{F}_q$ to $\mathbb{F}_2$. This should be large in order to be resistant to algebraic attacks, whose impact on breaking filtering generators have been recently shown in the literature [6] [7]. (*Note.* Using a lower degree polynomial to approximate a filtering sequence is not a new approach, which has been studied as a linearized method for NLFSR back in 1970s.)

There are considerably amount of publications for constructing boolean functions with some of these three properties. But none of those functions produce span n sequences. On the other hand, the corresponding de Bruijn sequence of a span n sequence always has a large linear span, at least $2^{n-1} + n$ (see Section 1). Thus, the genuine unpredictability of a de Bruijn sequence is determined by the unpredictability of the corresponding span n sequence. We define the linear span of the corresponding de Bruijn sequence of a span n sequence as an *illusional linear span* of the span n sequence. Let b be the corresponding de Bruijn sequence of a. In addition to the above randomness criteria F1-F3, we require that

(F4) The difference between the illusional linear span and the linear span of a span n sequence should be small, i.e., $\Delta = |LS(a) - LS(b)|$ is small or $LS(a) \geq 2^{n-1}$.

It is not easy to determine whether a span n sequence, regarded as a filtering sequence, satisfies the randomness properties F1-F4 as well as possessing some other randomness properties. Up to now, there are no sub-classes of span n sequences whose feedback functions are algebraically known except for those constructed from m-sequences by adding one zero into the run of zeros of length $n-1$. However, these randomness criteria could be served as guidelines for future research in theory and test beds for design of secure communication systems in practice.

3.2   Examples

Example 1. Let

$$a = 0111110111001010011010110001000$$

Then a is a span 5 sequence. Let $\mathbb{F}_{2^5}$ be defined by a primitive polynomial $t(x) = x^5 + x^4 + x^3 + x + 1$, and α be a root of $t(x)$. Using the DFT, we can obtain the trace representation of a as follows

$$f(x) = Tr(x + x^3 + \alpha x^5 + \alpha^{22} x^7 + \alpha^{18} x^{11} + \alpha^7 x^{15}).$$

Thus $LS(a) = 30$. The distributions of the autocorrelation, additive autocorrelation, and Hadamard transform are given as follows.

Autocorrelation					
$C_a(\tau)$	31	-1	-5	7	-9
#$\tau's$	1	16	4	6	4

Hadamard Transform						
$\widehat{f}(\lambda)$	0	4	-4	8	-8	12	-12
#$\lambda's$	8	7	7	6	2	1	1

Additive Autocorrelation			
$A_f(\omega)$	0	8	-8	-16
#$\omega's$	16	7	7	2

Thus, we have the nonlinearity or linear resistancy of a: $N_a = 10$, which is smaller than the maximum linear resistancy 12 for $n = 5$, and the differential resistancy of a: $D_a = 8$, which is poor, compared to the maximum differential resistancy 12 for $n = 5$. The illusional linear span of a is equal to 27. Therefore $\Delta = 3$. Thus, this sequence has good linear span property whenever we consider the difference between a and b or the linear span of a solely.

Remark 1. The above randomness criteria F1-F4 are defined for span n sequences. However, they could apply to any binary sequences of period N. For n odd, both the optimal linear resistancy and optimal differential resistancy are given by $A = 2^{n-1} - 2^{(n-1)/2}$. For $n = 5$ and 7, $N_a = A$ is maximum. For the differential resistancy, no examples have been found which satisfies $D_a > A$.

In the following example, we show randomness of a class of span n sequences constructed by Golomb in [13].

Example 2. Let $f(x) = x^n + x + 1$, n odd, which is primitive over $\mathbb{F}_2$, let $1\underbrace{00\cdots0}_{n-1}$ be an initial state of the LFSR with $f(x)$ as the minimal polynomial which generates a. Then we have

$$a_{N-1}, a_0, \cdots, a_{n-1}, a_n, a_{n+1}, \cdots, a_{2n-1} = 11\underbrace{00\cdots0}_{n-1}1\underbrace{00\cdots0}_{n-2}1$$

In [13], Golomb constructed a class of span n sequences with the constant-on-cosets property for n odd as follows. (We say a sequence $a = \{a_i\}$ satisfies the constant-on-cosets property if there exists a k-shift of a such that $a_{k+2i} = a_{k+i}, i = 0, 1, \cdots$.) Let $b_i = a_i + 1$, $i = 1, \cdots, N-1$ and $b_0 = a_0$. Then b is a span n sequence with the constant-on-coset property. We notice that b is obtained by complementing every bit of a except for a_0. In the following, we show that the linear span of b is equal to $2^n - 2 - n$. Assume that $a_i = Tr(\beta\alpha^i), i = 0, 1, \cdots, n-1, \beta \in \mathbb{F}_{2^n}$ where α is a root of $x^n + x + 1$. Let $\{A_j\}$ and $\{B_j\}$ be the spectral sequences of a and b respectively. Since $\{a_i\}$ satisfies the constant-on-cosets property, then $\beta = 1$ (see [15] for this result). From the DFT, we have

$$B_j = \sum_{i=0}^{N-1} b_i\alpha^{-ij} = a_0 + \sum_{i=1}^{N-1}(a_i + 1)\alpha^{-ij}$$

$$= \sum_{i=1}^{N-1}\alpha^{-ij} + \sum_{i=0}^{N-1} a_i\alpha^{-ij} = 1 + A_j.$$

Since a is an m-sequence, then $A_j = 0$ for all the coset leaders $j \neq 1$. Note that $A_1 = \beta$, and $\beta = 1$. Thus $LS(b) = 2^n - 2 - n$, which is very large, compared with the maximum linear span $2^n - 2$. However, the corresponding filtering function of b has very poor nonlinearity, which is demonstrated as follows. Let $f(x)$ and $g(x)$ be their respective trace representations of a and b. Then $f(x) = Tr(x)$ and

$$g(x) = \begin{cases} 1 + Tr(x) & \text{if } x \neq 0, 1 \\ 1 & \text{if } x = 1 \\ 0 & \text{if } x = 0. \end{cases}$$

Therefore, the Hadamard transform of $g(x)$ is computed by

$$\hat{g}(\lambda) = \sum_{x \in \mathbb{F}_{2^n}} (-1)^{Tr(\lambda x) + g(x)}$$

$$= 2 - 2(-1)^{Tr(\lambda)} - \sum_{x \in \mathbb{F}_{2^n}} (-1)^{Tr((\lambda+1)x)}.$$

Note that n is odd, so $Tr(1) = 1$. By examining the above summations, we have $\hat{g}(\lambda) \in \{0, 4, 4 - 2^n\}$. According to the definition of the nonlinearity, we have $c_g = 2^n - 4$ and $N_f = 2$, i.e., the nonlinearity of g is equal to 2, which is almost the worst nonlinearity of a nonlinear function.

Remark 2. The poor randomness of b can also be detected from the 1-error linear span of $\bar{b}$, the complement of b. (*Note.* The k-error linear span of a sequence u of period r is defined by $E_k(u) = \min\{LS(u + e) \mid e \in \mathbb{F}_2^r, H(e) = k\}$ where $H(e)$ is the Hamming weight of the r-dimensional vector e, and $LS(x)$ is the linear span of x.) According to the construction of b, we have $\bar{b} = a + e$ where $e = (1, 0, \cdots, 0)$. Thus $E_1(\bar{b}) = E_1(a + e) = n$. From this observation, it is worth to look at k-error linear spans of sequences through their nonlinearity, since the latter have received a tremendous large amount of publications in recent 30 years.

4   Relationships Between DFTs and Feedback Functions

In this section, we investigate how feedback functions of span n sequences can be derived from their trace representations, i.e., their DFTs, and vice versa. Let $a = \{a_k\}$ be a span n sequence, and $f(x)$ be its trace representation. Let $S_k = (a_k, \cdots, a_{k+n-1})$ be the kth state of a, $\mathcal{S} = \{S_k \mid 0 \le k < N\}$, and $\mathbb{F}_q^*$, the multiplicative group of $\mathbb{F}_q$ which are represented by powers of α, a primitive element in $\mathbb{F}_q$. Let

$$\sigma : S_k \to \alpha^k, 0 \le k < N, \text{ or equivalently}$$

$$\sigma : (a_k, a_{k+1}, \cdots, a_{k+n-1}) \to (c_{k,0}, c_{k,1}, \cdots, c_{k,n-1}) \tag{7}$$

where $\alpha^k = \sum_{j=0}^{n-1} c_{k,j} \alpha^j$. This map induces a permutation on $\mathbb{Z}_N$ in the following fashion:

$$\pi = \begin{pmatrix} 0 & 1 & \cdots & k & \cdots & N-1 \\ t_0 & t_1 & \cdots & t_k & \cdots & t_{N-1} \end{pmatrix} \tag{8}$$

where t_k is determined by $\sum_{j=0}^{n-1} a_{k+j} \alpha^j = \alpha^{t_k}$. Since σ is a one-to-one map between $\mathcal{S}$ and $\mathbb{F}_q^*$, then σ^{-1} exists. Therefore, we may write

$$a_k = f(\alpha^k) = f(\sigma\sigma^{-1}(\alpha^k)) = f(\sigma(S_k)), k = 0, 1, \cdots. \tag{9}$$

We define $b = \{b_k\}$, and $b_{n+k} = g(S_k), k = 0, 1, \cdots$ where $g(x) = f \circ \sigma(x)$, the composition of f and σ, as a feedback function of an NLFSR, and an initial state of b is given by $(b_0, \cdots, b_{n-1}) = (a_{N-n}, \cdots, a_{N-1})$. According to (1) in Section 2, b is generated by the NLFSR with n stages and the feedback function g. From (9), we have $a = L^n(b)$. Thus, we have established the following theorem.

Theorem 1. *With the above notation, then $g = f \circ \sigma$ is the feedback function of an NLFSR which generates b, an $(N - n)$-shift of a, i.e., $b = L^{N-n}(a)$. So, they have equal linear spans.*

Given that $f(x)$ is a feedback function of an NLFSR which generates a span n sequence a, we would like to ask what the trace representation of a is. Again, using (1),

$$a_{n+k} = f(S_k), k = 0, 1, \cdots, \text{ where}$$

$$S_k = (a_k, a_{k+1}, \cdots, a_{k+n-1}). \tag{10}$$

We now define the same map as defined in (7). From (10),

$$a_{n+k} = f(\sigma^{-1}\sigma(S_k)) = f(\sigma^{-1}(\alpha^k)), k = 0, 1, \cdots. \tag{11}$$

We denote $h(x) = f \circ \sigma^{-1}(x)$, and $c = \{c_k\}$ where $c_k = h(\alpha^k), k = 0, 1, \cdots$. Then $h(x)$ is the trace representation of c. However, from (11), $c = L^n(a)$, so that $a = L^{N-n}(c)$. Let $\{A_k\}$ and $\{C_k\}$ be the spectral sequences of a and c respectively. Applying Fact 2, we have $A_k = \alpha^{-nk}C_k$. Thus, we have proved the inverse of Theorem 1, which is shown below.

Theorem 2. *Let $f(x)$ be a feedback function of an NLFSR which generates a span n sequence a. Then the trace representation of a is given by $f \circ \sigma^{-1}(\alpha^{-n}x)$.*

For an easy comparison of these relationships, we summarize the results of Theorems 1 and 2 into the following table.

Relationships Between DFT and Feedback Functions

DFT	Feedback Function in an NLFSR
$a \leftrightarrow f(x)$	$f \circ \sigma(x)$ generates $L^{N-n}(a)$
$a \leftrightarrow f \circ \sigma^{-1}(\alpha^{-n}x)$	f generates a

In the following example, a feedback function is a linear function, since there are no span 3 sequences which are not m-sequences. This is only for demonstrating the principle of these two theorems.

Example 3. Let $n = 3$, and α be a primitive element in $\mathbb{F}_{2^3}$ with $\alpha^3 + \alpha + 1 = 0$. Let $f(x_0, x_1, x_2) = x_0 + x_1$, and let $(a_0, a_1, a_2) = (0, 0, 1)$ be an initial state of the LFSR. Then $a = 0010111$, an m sequence of period 7. Using the DFT (see Section 2), the trace representation of a is equal to $Tr(\alpha x)$. In the following, we use the method in Theorem 2 to obtain the trace representation of a. We compute the polynomial form $f(x_0, x_1, x_2) = Tr(\alpha^6 x)$. The map σ is given by the following table.

t_k	$\sigma : S_k \rightarrow$	α^k, exponents k
2	001	0
1	010	1
6	101	2
4	011	3
5	111	4
3	110	5
0	100	6

Using the DFT of functions (or the Lagrange interpolation), we compute that $\sigma^{-1}(x) = \alpha^4 x + x^2 + \alpha^3 x^4$. Thus

$$f \circ \sigma^{-1}(x) = Tr(\alpha^6(\alpha^4 x + x^2 + \alpha^3 x^4)) = Tr(\alpha^4 x).$$

According to Theorem 2, the trace representation of a is given by $f \circ \sigma^{-1}(\alpha^{-3}x) = Tr(\alpha x)$, which verifies the result computed directly from the DFT of a.

Remark 3. The power of Theorems 1 and 2 is a new look at the span n sequences. These results present specific relationships between a feedback function in an NLFSR which generates a span n sequence and the trace representation of the sequence through the one-to-one map from its state space to the multiplicative group of $\mathbb{F}_q$. On the other hand, this provides an algebraic way to construct span n sequences, which is different from all the known constructions for de Bruijn sequences. For example, one could find all span 7 sequences with linear span 21 by testing all the sums of three m-sequences of period 127. In other words, one does the search for span n sequences with trace representations of $Tr(\gamma_1 x^r + \gamma_2 x^s + \gamma_3 x^t)$ where γ_i's run through $\mathbb{F}_q^*$ and r, s and t are distinct coset leaders modulo $2^n - 1$. Since there are 18 different cosets modulo 127, the search complexity is of $18 \times 17 \times 16 \times 127^2 = 78967584$ trials. (Note a shifted sequence of a span n sequence preserves the span n property. Thus the choices of γ_i are reduced from 127^3 down to 127^2.) So, the corresponding de Bruijn sequences of period 128 are obtained.

Remark 4. For an arbitrary NLFSR, if the period of an output sequence, say r, divides N, then the results are similar as those of Theorems 1 and 2 in which the primitive element α is replaced by an element α in $\mathbb{F}_q$ with order r. However, if r is not a divisor of N, then such relationships do not exist.

Acknowledgment. The author wishes to thank the referees for their valuable and helpful comments. The research is supported by NSERC Discovery Grant and SPG.

References

1. Beker, H., Piper, F.: Cipher Systems. John Wiley and Sons, New York (1982)
2. Berlekamp, E.R.: Algebraic coding theory. McGraw-Hill, New York (1968)
3. Biham, E., Shamir, A.: Differential Cryptanalysis of DES-like Cryptosystems. In: Menezes, A.J., Vanstone, S.A. (eds.) CRYPTO 1990. LNCS, vol. 537, pp. 2–21. Springer, Heidelberg (1991)
4. Chan, A.H., Games, R.A., Key, E.L.: On the complexities of de Bruijn sequences. J. Combin. Theory 33, 233–246 (1982)
5. Chabaud, F., Vaudenay, S.: Links between differential and linear cryptanalysis. In: De Santis, A. (ed.) EUROCRYPT 1994. LNCS, vol. 950, pp. 363–374. Springer, Heidelberg (1995)
6. Courtois, N., Meier, W.: Algebraic attacks on stream ciphers with linear feedback. In: Biham, E. (ed.) Advances in Cryptology – EUROCRPYT 2003. LNCS, vol. 2656, pp. 345–359. Springer, Heidelberg (2003)
7. Courtois, N.: Fast algebraic attacks on stream ciphers with linear feedback. In: Boneh, D. (ed.) CRYPTO 2003. LNCS, vol. 2729, pp. 176–194. Springer, Heidelberg (2003)
8. Ding, C., Xiao, G., Shan, W.: The Stability Theory of Stream Ciphers. LNCS, vol. 561. Springer, Heidelberg (1991)

9. Etzion, T., Lempel, A.: Construction of de Bruijn sequences of minimal complexity. IEEE Trans. Inform. Theory IT-30(5), 705–709 (1984)
10. eSTREAM - The ECRYPT Stream Cipher Project, http://www.ecrypt.eu.org/stream
11. Goldreich, O.: Foundations of Cryptography: Basic Applications. Cambridge University Press, Cambridge (2004)
12. Golomb, S.W.: Shift Register Sequences. Holden-Day, Inc., San Francisco (1967); revised edition, Aegean Park Press, Laguna Hills, CA (1982)
13. Golomb, S.W.: On the classification of balanced binary sequences of period $2^n - 1$. IEEE Trans. on Inform. Theory IT-26(6), 730–732 (1980)
14. Golomb, S.W.: Irreducible polynomials, synchronization codes, primitive necklaces, and the cyclotomic algebra. In: Bose, R.C., Dowling, T.A. (eds.) Combinatorial Mathematics and its Applications, pp. 358–370. University of North Carolina Press, Chapel Hill (1969)
15. Golomb, S.W., Gong, G.: Signal Design with Good Correlation: for Wireless Communications, Cryptography and Radar Applications. Cambridge University Press, Cambridge (2005)
16. National Institute of Standards and Technology, Digital Signature Standard (DSS), Federal Information Processing Standards Publication, FIPS PUB 186-2, Reaffirmed (January 27, 2000)
17. Mayhew, G.L., Golomb, S.W.: Linear spans of modified de Bruijn sequences. IEEE Trans. Inform. Theory IT-36(5), 1166–1167 (1990)
18. Massey, J.L.: Shift-register synthesis and BCH decoding. IEEE Trans. on Inform. Theory 15(1), 81–92 (1969)
19. Matsui, M.: Linear cryptanalysis method for DES cipher. In: Helleseth, T. (ed.) EUROCRYPT 1993. LNCS, vol. 765, pp. 386–397. Springer, Heidelberg (1994)
20. Menezes, A.J., van Oorschot, P.C., Vanstone, S.A.: Handbook of Applied Cryptography. CRC Press, Boca Raton, USA (1996)
21. Rueppel, R.A.: Analysis and Design of Stream Ciphers. Springer, Heidelberg (1986)
22. Shannon, C.E.: Communication Theory of Secrecy Systems. Bell System Technical Journal XXVII (4), 656–715 (1949)
23. Siegenthaler, T.: Correlation-immunity of nonlinear combining functions for cryptographic applications. IEEE Trans. Inform. Theory 30(5), 776–780 (1984)

On Attacks on Filtering Generators Using Linear Subspace Structures

Sondre Rønjom[1], Guang Gong[2], and Tor Helleseth[1]

[1] The Selmer Center,
Department of Informatics, University of Bergen
PB 7803
N-5020 Bergen, Norway
[2] Department of Electrical and Computer Engineering
University of Waterloo
Waterloo, Ontario N2L 3G1, Canada

Abstract. The filter generator consists of a linear feedback shift register (LFSR) and a Boolean filtering function that combines some bits from the shift register to create a key stream. A new attack on the filter generator has recently been described by Rønjom and Helleseth [6]. This paper gives an alternative and extended attack to reconstruct the initial state of the LFSR using the underlying subspace structure of the filter sequence. This improved attack provides further insight and more flexibility in performing the attack by Rønjom and Helleseth. The main improvements are that this attack does not use the coefficient sequences that were fundamental in the previous attack and also works in the unlikely cases when the original attack needed some modifications.

Keywords: filter generator, m-sequences, stream ciphers.

1 Introduction

The filter generator uses a primitive linear feedback shift register(LFSR) of length n that generates a maximal linear sequence (an m-sequence) $\{s_t\}$ of period $2^n - 1$ satisfying a recursion with a characteristic polynomial $h(x) \in \mathbb{F}_2[x]$ of degree n being a primitive polynomial with primitive zeroes α^{2^i} for $i = 0, 1, \ldots, n - 1$.

At each time t, a key stream bit b_t is calculated as a function of certain bits in some positions $(e_0, e_1, \ldots, e_{m-1})$ in the LFSR state $(s_t, s_{t+1}, \ldots, s_{t+n-1})$ at time t using a Boolean function $f(x_0, x_1, \ldots, x_{m-1})$ of degree d in $m \leq n$ variables. The key stream is defined by

$$b_t = f(s_{e_0+t}, s_{e_1+t}, \ldots, s_{e_{m-1}+t}).$$

Since s_t is a linear combination of the bits in the initial state $(s_0, s_1, \ldots, s_{n-1})$ this leads to an equation system

$$b_t = f_t(s_0, s_1, \ldots, s_{n-1}) \text{ for } t = 0, 1, \ldots$$

which has the initial state of the LFSR as a solution.

S.W. Golomb et al. (Eds.): SSC 2007, LNCS 4893, pp. 204–217, 2007.

In a recent paper Rønjom and Helleseth [6] present a new attack that reconstructs the initial state $(s_0, s_1, \ldots, s_{n-1})$ of the binary filter generator using D key stream bits with complexity $O(D)$, where $D = \sum_{i=1}^{d} \binom{n}{i}$, after a precomputation of complexity $O(D(log_2D)^3)$. If L is the linear complexity of the key stream, and we have enough information about the zeros of the minimal polynomial of $\{b_t\}$, then D can be replaced by L in these complexity estimates.

The underlying method in [6] is based on the observation that since

$$s_t = \sum_{i=0}^{n-1} (k_i\alpha^t)^{2^i} = Tr(k_0\alpha^t)$$

the key stream $\{b_t\}$ can be generated by a characteristic polynomial $p(x)$ with zeroes among α^J, where the Hamming weight of the binary representation of J, denoted $wt(J)$, obey $1 \leq wt(J) \leq d = deg(f)$. Therefore, we have

$$b_t = \sum_{\beta} d_\beta \beta^t$$

where $p(\beta) = 0$ and $d_\beta \in \mathbb{F}_{2^n}$. This polynomial can be constructed in the precomputation phase with complexity $O(D(log_2D)^3)([3])$.

The attack in [6] selects an irreducible polynomial $k(x)$ of degree n (in [6] the LFSR polynomial $h(x)$ was selected) and define the polynomial $p^*(x) = \frac{p(x)}{k(x)}$. The authors apply the shift operator $p^*(x) = \sum_{j=0}^{D-n} p_jx^j$ to the key stream $\{b_t\}$ i.e., compute

$$\sum_{j=0}^{D-n} p_jb_{t+j} = \sum_{j=0}^{D-n} p_jf_{t+j}(s_0, s_1, \ldots, s_{n-1})$$

and show that this almost always leads to a nonsingular linear equation system in the unknowns $s_0, s_1, \ldots, s_{n-1}$. The main reason for this is the simple observation that

$$\sum_{j=0}^{D-n} p_jb_{t+j} = \sum_{\beta} d_\beta p^*(\beta)\beta^t$$

where the summation now is only over the n zeros of $k(x)$ which implies that the right hand side for the case when $k(x) = h(x)$ can be written as $Tr(d_\alpha\alpha^t)$, a linear combination of bits in the initial state. In the very unlikely case that the system is trivial which happens when $d_\alpha = 0$ one needs to do some modifications. In [6], the irreducible polynomial $h(x)$ was suggested for $k(x)$, while a better choice in this case would be, for example, to use a different irreducible polynomial for $k(x)$. This will be described in this paper.

In [8], the authors view the filter generator entirely in terms of powers of a unique linear operator T together with a vector representing the filtering function. It is proved that T embodies the coefficient sequences described in [6]. Thus, properties of the vector space generated by the filter generator (for instance its dimension) can be determined using theory of cyclic vector spaces, which very elegantly complements analysis in terms of the roots of the LFSR.

In this paper we will take advantage of the fact that the filtering sequence can be written in a unique way as a sum of sequences with characteristic polynomials being all irreducible divisors of $p(x)$. Using a suitable shift operator to the key stream we can find any component sequence and find relations that determine the initial state of the LFSR for any key stream $\{b_t\}$. Since this paper only needs the linear subspace structure of the filter generator, the results can be directly extended to a combination generator as well. Although the technique presented here resembles the the technique in [6], it also covers some cases occurring with a small probability where the original technique would need some modifications.

2   Notations and Preliminaries

For more general treatments on minimal polynomials of the elements in a finite field or a periodic sequence, and the DFT of sequences, the reader is referred to [1].

A. The Left Shift Operator

Let $q = p^h$ for a prime integer p and a positive integer h. We denote a finite field of q elements by $\mathbb{F}_q$. Let $V(\mathbb{F}_q)$ be a set consisting of all infinite sequences whose elements are taken from $\mathbb{F}_q$, i.e.,

$$V(\mathbb{F}_q) = \{\mathbf{s} = (s_0, s_1, \cdots) \mid s_t \in \mathbb{F}_q\}.$$

Then $V(\mathbb{F}_q)$ is a linear space over $\mathbb{F}_q$. Let $\mathbf{s} = (s_0, s_1, s_2, \cdots) \in V(\mathbb{F}_q)$ whose elements satisfy the linear recursive relation

$$s_{t+n} = \sum_{i=0}^{n-1} h_i s_{t+i}, t = 0, 1, \cdots$$

Here $\mathbf{s}$ is referred to as a linear recursive sequence or it is a linear feedback shift register sequence generated by $h(x) = x^n - (h_{n-1}x^{n-1} + \cdots + h_0)$, and $h(x)$ is called a characteristic polynomial of $\mathbf{s}$. Furthermore, the characteristic polynomial of $\mathbf{s}$ with the smallest degree is called the minimal polynomial of $\mathbf{s}$.

The (left) shift operator E is defined as follows:

$$E\mathbf{s} = (s_1, s_2, s_3, \cdots) \text{ and } E^t\mathbf{s} = (s_t, s_{t+1}, s_{t+2}, \cdots), t \geq 1.$$

By convention, we write $E^0\mathbf{s} = I\mathbf{s} = \mathbf{s}$, where I is the identity transformation on $V(\mathbb{F}_q)$. Then $h(E)\mathbf{s} = \mathbf{0}$ where $\mathbf{0} = (0, 0, \cdots)$ is the zero sequence in $V(\mathbb{F}_q)$. For any non-zero polynomial $h(x) \in \mathbb{F}_q[x]$, we use $G(h)$ to represent the set consisting of all sequences in $V(\mathbb{F}_q)$ with

$$h(E)\mathbf{s} = \mathbf{0}.$$

Since $h(E)$ is also a linear transformation, $G(h)$ is a subspace of $V(\mathbb{F}_q)$.

B. m-sequences and Trace Representation

Let $h(x)$ be a primitive polynomial of degree n, and let α be a root of $h(x)$ in $\mathbb{F}_{q^n}$. Then the trace representation of $\mathbf{s}$ is given by

$$s_t = Tr(\beta \alpha^t), t = 0, 1, \ldots, \beta \in \mathbb{F}_{q^n}$$

where $Tr(x)$ denotes the trace function $Tr(x) = \sum_{i=0}^{n-1} x^{q^i}$ which is a map from $\mathbb{F}_{q^n}$ to $\mathbb{F}_q$.

C. Filter Generators

Select $m \le n$ integers: $0 \le e_0 \le e_1 \le \ldots \le e_{m-1} < n$, and a polynomial function $f(x_0, x_1, \ldots, x_{m-1}) \in \lambda = \mathbb{F}_q[x_0, x_1, \ldots, x_{m-1}]/(x_i^2 + x_i)_{0 \le i < m}$ of degree d that can be written in the form

$$f(x_0, x_1, \cdots, x_{m-1}) = \sum c_{i_0, i_1, \cdots, i_{m-1}} x_{i_0}^{i_0} x_{i_1}^{i_1} \cdots x_{i_{m-1}}^{i_{m-1}}, \ c_{i_0, i_1, \cdots, i_{m-1}} \in \mathbb{F}_q.$$

A sequence $\mathbf{b} = \{b_t\}$ whose elements are defined by a function of

$$b_t = f(s_{e_0+t}, s_{e_1+t}, \ldots, s_{e_{m-1}+t}) = f_t(s_0, s_1, \ldots, s_{n-1}), t = 0, 1, \ldots,$$

is called a filtering sequence and the function $f(x_0, x_1, \ldots, x_{m-1})$ is called a filtering function.

D. DFT and Inverse DFT

Let $\{a_t\}$ be a sequence over $\mathbb{F}_q$ with period $N = q^n - 1$. Recall that α is a primitive element in $\mathbb{F}_{q^n}$. Then the *(discrete) Fourier Transform (DFT)* of $\{a_t\}$ is defined by

$$A_k = \sum_{t=0}^{N-1} a_t \alpha^{-tk}, \ k = 0, 1, \ldots, N-1.$$

The inverse DFT (IDFT) is given by

$$a_t = \frac{1}{N} \sum_{k=0}^{N-1} A_k \alpha^{kt}, \ t = 0, 1, \ldots, N-1.$$

The sequence $\{A_k\}$ is referred to as a *spectral sequence* of $\{a_t\}$. Let $A(x) = \sum_{k=0}^{N-1} A_k x^k$. Then $A(x)$ can be written as

$$A(x) = \sum_k Tr_1^{m_k}(A_k x^k)$$

where the k's are (cyclotomic) coset leaders modulo N, and $m_k \mid n$ is the length of the coset which contains k. This is called a *trace representation* of $\{a_t\}$.

3   Minimal Polynomials with Constrained Weights

In this section, we show the linear subspace structure of the filtering sequence. For a positive integer i, we define

$$H(i) = \sum_{j=0}^{n-1} i_j,$$

where

$$i = \sum_{j=0}^{n-1} i_j q^j, \ 0 \le i_j < q,$$

which is referred to as the weight of i. Note that $H(i)$ is the usual Hamming weight of i when $q = 2$. It is known that the zeroes of the minimal polynomial of the filtering sequence $\mathbf{b}$ is a subset of the following set:

$$\Omega(d) = \{\alpha^i \mid H(i) \le d\} \tag{1}$$

where $d = deg(f) \ (\le m)$.

Let $g_{\alpha^i}(x)$ be the minimal polynomial of α^i over $\mathbb{F}_q$ which is given by

$$g_{\alpha^i}(x) = \prod_{j=0}^{n_i-1} (x - \alpha^{i \cdot q^j}),$$

where n_i is the size of the coset C_i which is the smallest number satisfying $i \equiv i \cdot q^{n_i} \pmod{q^n - 1}$. Let T be the set consisting of all coset leaders modulo $q^n - 1$ and define

$$T(d) = \{i \in T \mid 1 \le H(i) \le d\}.$$

Let $p(x)$ be the polynomial

$$p(x) = \prod_{i \in T(d)} g_{\alpha^i}(x)$$

and $p_i(x)$ the polynomial

$$p_i(x) = \frac{p(x)}{g_{\alpha^i}(x)}.$$

Then $G(p)$ can be written as a direct sum of the subspaces of $V(\mathbb{F}_q)$

$$G(p) = G(g_{\alpha^{t_1}}) \oplus \ldots \oplus G(g_{\alpha^{t_s}})$$

where $t_j \in T(d) = \{t_1, \ldots, t_s\}$. Let $\mathbf{a}_k = \{a_{k,t}\}_{t \ge 0} \in G(g_{\alpha^k})$ where $k \in T(d)$. Then we have that

$$b_t = \sum_{k \in T(d)} a_{k,t}. \tag{2}$$

4   Extractors

Let L be the linear span of $\{b_t\}$, and $l = D - n$. From (1), L is upper bounded by the cardinality of $\Omega(d)$, which is $D = \sum_{i=1}^{d} \binom{n}{i}$ for $q = 2$, the binary case. We may write

$$b_t = \sum_{k \in T(d)} Tr_1^{n_k}(A_k(\beta\alpha^t)^k), t = 0, 1, \dots$$

and

$$a_{k,t} = Tr_1^{n_k}(A_k(\beta\alpha^t)^k), t = 0, 1, \dots.$$

For k with $gcd(k, q^n - 1) = 1$, let $p_k(x) = x^l + \sum_{i=0}^{l-1} c_i x^i, c_i \in \mathbb{F}_q$. For avoiding the use of a double index, we denote

$$\mathbf{a}_k = \{a_{k,t}\}_{t \geq 0} = \{a_t\}_{t \geq 0}.$$

Then we have

$$a_t = Tr(A_k(\beta\alpha^t)^k), t = 0, 1, \dots.$$

In the following, we show how to separate or extract $\mathbf{a}_k$ from $\mathbf{b} = \{b_t\}$, the filtering sequence. From (2), it follows that

$$\mathbf{b} = \sum_{j=1}^{s} \mathbf{a}_j$$

and thus we have that

$$p_k(E)\mathbf{b} = \sum_{j=1}^{s} p_k(E)\mathbf{a}_j. \tag{3}$$

Note that $p_k(E)\mathbf{a}_j = 0$ if $j \neq k$. Thus (3) above becomes

$$p_k(E)\mathbf{b} = p_k(E)\mathbf{a}_k. \tag{4}$$

If we let

$$u_t = a_{l+t} + \sum_{i=0}^{l-1} c_i a_{i+t}, t = 0, 1, \dots,$$

we see that

$$p_k(E)\mathbf{a}_k = (u_0, u_1, \dots).$$

Going through the details, we have that

$$u_t = a_{l+t} + \sum_{i=0}^{l-1} c_i a_{i+t}$$

$$= Tr(A_k(\beta\alpha^{l+t})^k) + \sum_{i=0}^{l-1} c_i Tr(A_k(\beta\alpha^{i+t})^k)$$

$$= Tr(A_k\beta^k(\alpha^{lk} + \sum_{i=0}^{l-1} c_i\alpha^{ik})\alpha^{tk})$$

$$= Tr(A_k\beta^k p_k(\alpha^k)\alpha^{tk})$$

$$= Tr(r\alpha^{tk})$$

where $r = A_k\beta^k p_k(\alpha^k)$ and $p_k(\alpha^k) \neq 0$ since α^k is not a root of $p_k(x)$. In general, we have therefore shown that

$$u_t = Tr(r\alpha^{tk}), t = 0, 1, \ldots, \tag{5}$$

where $r = A_k\beta^k p_k(\alpha^k)$.

Let $g(x)$ be the minimal polynomial of $\mathbf{b} = \{b_t\}$ and $gcd(k, 2^n - 1) = 1$. It follows that

$$g_{\alpha^k}(x)|g(x) \Leftrightarrow (u_0, \ldots, u_{n-1}) \neq 0$$
$$\Leftrightarrow A_k \neq 0.$$

We call $(u_0, u_1, \cdots, u_{n-1})$ an *extractor* of $\mathbf{b}$.

5   Extract β

Let $(b_0, b_1, \ldots, b_{D-1})$ be known (or sometimes $(b_0, b_1, \ldots, b_{L-1})$ may be sufficient). The goal is to obtain β. This yields the initial state of the LFSR which produces $\mathbf{b}$. From (4) we have that

$$p_k(E)\mathbf{b} = p_k(E)\mathbf{a}_k = (u_0, u_1, \ldots)$$

and so

$$u_0 = \sum_{i=0}^{l} c_i b_i$$

$$u_1 = \sum_{i=0}^{l} c_i b_{i+1}$$

$$\vdots$$

$$u_{n-1} = \sum_{i=0}^{l} c_i b_{i+n-1}.$$

Thus $(u_0, u_1, \ldots, u_{n-1})$ can be computed from $(b_0, b_1, \ldots, b_{D-1})$. From (5) and $(u_0, \ldots, u_{n-1})$ a system of equations with unknown β is formed

$$u_0 = \quad Tr(r) = \quad r + r^2 + \ldots + r^{2^{n-1}}$$

$$u_1 = \quad Tr(r\gamma) = \quad \gamma r + \gamma^2 r^2 + \ldots + \gamma^{2^{n-1}} r^{2^{n-1}}$$

$$\vdots$$

$$u_{n-1} = Tr(r\gamma^{n-1}) = \gamma^{n-1} r + \gamma^{(n-1)2} r^2 + \ldots + \gamma^{(n-1)2^{n-1}} r^{2^{n-1}}$$

where $\gamma = \alpha^k$.

Let $x_i = r^{2^i}$ and $\alpha_i = \alpha^{k2^i}$ for $i = 0, 1, \ldots, n-1$ and form a matrix M of the form

$$M = \begin{bmatrix} 1 & 1 & \cdots & 1 \\ \alpha_0 & \alpha_1 & \cdots & \alpha_{n-1} \\ \alpha_0^2 & \alpha_1^2 & \cdots & \alpha_{n-1}^2 \\ \vdots & \vdots & & \vdots \\ \alpha_0^{n-1} & \alpha_1^{n-1} & \cdots & \alpha_{n-1}^{n-1} \end{bmatrix}.$$

Then we have

$$M \begin{bmatrix} x_0 \\ x_1 \\ \vdots \\ x_{n-1} \end{bmatrix} = \begin{bmatrix} u_0 \\ u_1 \\ \vdots \\ u_{n-1} \end{bmatrix}.$$

Since M is a Vandermonde matrix and $(u_0, u_1, \ldots, u_{n-1})$ is known, by solving this equation system we obtain $x_0 = r$. This gives

$$r = A_k \beta^k p_k(\alpha^k)$$

and therefore

$$\beta^k = r A_k^{-1} [p_k(\alpha^k)]^{-1}$$

where r and $p_k(\alpha^k)$ are known. The remaining task is how to find A_k. Note that $\{A_k\}$ is related to a discrete Fourier transform of $\{b_t\}$, which can be computed through expansion of b_t.

6   How to Compute $\{A_k\}$

In this section, we restrict ourselves to the binary case. For the q-ary $(q > 2)$ case, there is a similar result which is omitted here for simplicity. In the binary case, $f(x_0, x_1, \ldots, x_{m-1}) = \sum_{(i_1, \cdots, i_e)} c_{i_1, \ldots, i_e} x_{i_1} \cdots x_{i_e}$. Then

$$b_t = f(s_{e_0+t}, \ldots, s_{e_{m-1}+t}) = \sum_{\underline{i}} c_{i_1, \ldots, i_e} s_{i_1+t} \cdots s_{i_e+t}$$

where $\underline{i} = \{i_1, \ldots, i_e\} \subset \{e_0, \ldots, e_{m-1}\}$. Let

$$y_t = s_{i_1+t} s_{i_2+t} \cdots s_{i_e+t}, t = 0, 1, \ldots$$

be a typical term. Since $s_t = Tr(\beta \alpha^t)$, we expand y_t

$$y_t = Tr(\beta \alpha^{i_1+t}) Tr(\beta \alpha^{i_2+t}) \cdots Tr(\beta \alpha^{i_e+t})$$

$$= [\sum_{v_1=0}^{n-1} (\beta \alpha^{i_1+t})^{2^{v_1}}] \cdots [\sum_{v_e=0}^{n-1} (\beta \alpha^{i_e+t})^{2^{v_e}}]$$

$$= \prod_{j=1}^{e} (\sum_{v=0}^{n-1} \alpha^{i_j 2^v} x^{2^v})$$

$$= \sum_{v_1, v_2, \ldots, v_e} \alpha^{i_1 2^{v_1} + i_2 2^{v_2} + \cdots + i_e 2^{v_e}} x^{2^{v_1} + 2^{v_2} + \cdots + 2^{v_e}}$$

where $x = \beta\alpha^t$.

The following theorem is useful for simplifying the calculations of y_t.

Theorem 1. *We have,*

$$y_t = \sum_{k \in T(e)} Y_{\underline{i},k} x^k, \underline{i} = (i_1, \ldots, i_e)$$

where

$$Y_{\underline{i},k} = \sum_u \sum_J \alpha^{J(u)} \qquad (6)$$

where

$$J(u) = \sum_{j=1}^{v} 2^{t_j} h_j$$

and $u = (u_0, u_1, \ldots, u_{n-1})$ is a solution of

$$\begin{cases} \sum_{i=0}^{n-1} u_i 2^i \equiv k \pmod{2^n - 1} \\ \sum_{i=0}^{n-1} u_i = e, u_i \geq 0, \end{cases}$$

and where J is obtained by a partition of $\{i_1, \ldots, i_e\}$ into v parts for which the jth part has size u_{t_j} with

$$\{t_1, \ldots, t_v\} = \{i \mid u_i \neq 0\}, v \leq e$$

and h_j is the sum of elements in the jth part.

Remark. The inner sum of (6) can be described by the sum of the distinct permutation terms in the determinant of the following $e \times e$ matrix:

$$E_u = \begin{bmatrix} \overbrace{\alpha^{i_1 2^{t_1}} \cdots \alpha^{i_1 2^{t_1}}}^{u_{t_1}} & \overbrace{\alpha^{i_1 2^{t_2}} \cdots \alpha^{i_1 2^{t_2}}}^{u_{t_2}} & \cdots & \overbrace{\alpha^{i_1 2^{t_v}} \cdots \alpha^{i_1 2^{t_v}}}^{u_{t_v}} \\ \vdots & \vdots & & \vdots \\ \alpha^{i_e 2^{t_1}} \cdots \alpha^{i_e 2^{t_1}} & \alpha^{i_e 2^{t_2}} \cdots \alpha^{i_e 2^{t_2}} & \cdots & \alpha^{i_e 2^{t_v}} \cdots \alpha^{i_e 2^{t_v}} \end{bmatrix}.$$

In other words, $\sum_J \alpha^{J(u)}$ is equal to the sum of different permutation terms of E_u. When $u_{t_j} = 1$ for all $j = 1, \cdots, v$ ($v = e$ in this case), we have $\sum_J \alpha^{J(u)} = \det E_u$, the determinant of E_u.

Theorem 2. *For*

$$b_t = \sum_{k \in T(d)} Tr_1^{n_k}(A_k(\beta\alpha^t)^k),$$

then

$$A_k = \sum_{\underline{i}:c_{i_1}, \ldots, i_e \neq 0} Y_{\underline{i},k}$$

where $Y_{\underline{i},k}$ is given by Theorem 1.

The proofs of these two theorems including the q-ary case, $q > 2$, can be found in [2] or derived from the results in [4] and [5].

Example 1. Consider a filter generator consisting of an m-sequence $\mathbf{s} = \{s_t\}$ generated by $h(x) = x^4 + x + 1 \in \mathbb{F}_2$, where we can write $\{s_t\}$ in terms of the zeroes of $h(x)$ by

$$s_t = Tr(\beta \alpha^t), t = 0, 1, \ldots, \beta \in \mathbb{F}_{2^4},$$

filtered through the simple function

$$f(x_0, x_1, x_3) = x_0 x_1 x_3$$

with $deg(f) = 3$. Let $(e_0, e_1, e_2) = (0, 1, 3)$ be the tapping positions from the register such that the key stream sequence $\{b_t\}$ is given by

$$b_t = f_t(s_0, s_1, s_2, s_3) = f(s_{t+e_0}, s_{t+e_1}, s_{t+e_2}) = s_t s_{t+1} s_{t+3}.$$

In this example we assume that we observe

$$b_t = 0001000010\ldots$$

The cosets modulo 15 are

$$C_1 = \{1, 2, 4, 8\}$$
$$C_3 = \{3, 6, 12, 9\}$$
$$C_5 = \{5, 10\}$$
$$C_7 = \{7, 14, 13, 11\},$$

so the corresponding polynomials $g_{\alpha^i}(x)$ are

$$g_\alpha(x) = \prod_{i \in C_1} (x + \alpha^i) = x^4 + x + 1$$

$$g_{\alpha^3}(x) = \prod_{i \in C_3} (x + \alpha^i) = x^4 + x^3 + x^2 + x + 1$$

$$g_{\alpha^5}(x) = \prod_{i \in C_5} (x + \alpha^i) = x^2 + x + 1$$

$$g_{\alpha^7}(x) = \prod_{i \in C_7} (x + \alpha^i) = x^4 + x^3 + 1.$$

Thus in this case $p(x)$ is simply the polynomial

$$p(x) = \prod_{i \in T(3)} g_{\alpha^i}(x) = \sum_{i=0}^{14} x^i = \frac{x^{15} + 1}{x + 1}.$$

We can write b_t in terms of the zeroes of $g(x)$ as

$$b_t = \sum_{k \in T(3)} Tr_1^{n_k}(A_k x^k).$$

Using Theorem 1, we now compute $\{A_k\}$. For $k = 1$, we determine $u = (u_0, u_1, u_2, u_3)$ satisfying $\sum_{i=0}^{3} u_i 2^i \equiv 1 \pmod{15}$ and $\sum_{i=0}^{3} u_i = 3$. There is only one such u, which is $u = (0, 0, 2, 1)$, so we compute

$$A_1 = \alpha^{4e_0 + 4e_1 + 8e_2} + \alpha^{4e_0 + 8e_1 + 4e_2} + \alpha^{8e_0 + 4e_1 + 4e_2}$$
$$= \alpha^{13} + \alpha^5 + \alpha$$
$$= \alpha^{14}.$$

For $k = 3$, we have two possible values $u \in \{(0, 1, 0, 2), (3, 0, 0, 0)\}$ such that $\sum_{i=0}^{3} u_i 2^i \equiv 3 \pmod{15}$ and $\sum_{i=0}^{3} u_i = 3$. Both values of u lead to values of $J(u)$ being α^4, so

$$A_3 = \alpha^4 + \alpha^4 = 0.$$

For $k = 5$, we have two possible values of $u \in \{(1, 2, 0, 0), (0, 0, 1, 2)\}$, leading to values of $J(u)$ being α^3 and α^{12} respectively, and therefore

$$A_5 = \alpha^3 + \alpha^{12} = \alpha^{10}.$$

For $k = 7$, we find only one value of $u = (1, 1, 1, 0)$, leading to $J(u)$ being α^8, and thus

$$A_7 = \alpha^8.$$

Thus we can now write the sequence b_t as

$$b_t = Tr_1^4(\alpha^{14}x + \alpha^8 x^7) + Tr_1^2(\alpha^{10}x^5)$$

where $x = \beta\alpha^t, t = 0, 1, \ldots$. We have that the linear span of $\mathbf{b}$ is $L = 10$, which is equal to the degree of the polynomial $r(x) = g_\alpha(x)g_{\alpha^5}(x)g_{\alpha^7}(x)$. We may choose $k = 7$ and compute

$$p_7(x) = \frac{r(x)}{g_{\alpha^7}(x)}$$
$$= x^6 + x^5 + x^4 + x^3 + 1.$$

Then we have that $u_i = Tr(r\alpha^{7i})$ where

$$r = A_7\beta^7 p_7(\alpha^7)$$
$$= \beta^7 \alpha^8 \alpha^{14}$$
$$= \beta^7 \alpha^7.$$

Then from $b_t = 0001000010\ldots$, we compute $p_7(E)\mathbf{b}$ and obtain

$$u_t = b_{t+6} + b_{t+5} + b_{t+4} + b_{t+3} + b_t$$

for $0 \leq t \leq 3$ and get $(u_0, u_1, u_2, u_3) = (1, 0, 1, 0)$. Then we compute the matrix M defined by

$$M = \begin{bmatrix} 1 & 1 & 1 & 1 \\ \alpha^7 & \alpha^{14} & \alpha^{13} & \alpha^{11} \\ \alpha^{14} & \alpha^{13} & \alpha^{11} & \alpha^7 \\ \alpha^6 & \alpha^{12} & \alpha^9 & \alpha^3 \end{bmatrix}$$

and M_1 obtained by interchanging column 1 of M with u^T, and equals

$$M_1 = \begin{bmatrix} 1 & 1 & 1 & 1 \\ 0 & \alpha^{14} & \alpha^{13} & \alpha^{11} \\ 1 & \alpha^{13} & \alpha^{11} & \alpha^7 \\ 0 & \alpha^{12} & \alpha^9 & \alpha^3 \end{bmatrix}.$$

Then using M_1, M and u we compute $x_0 = r$ by

$$x_0 = \frac{\det M_1}{\det M} = \alpha^{13},$$

and since we also have $x_0 = r = \beta^7 \alpha^7$, we are left with solving for β and get $\beta = \alpha^3$. So the initial state is found to be

$$\begin{aligned} (s_0, s_1, s_2, s_3) &= (Tr(\beta), Tr(\beta\alpha), Tr(\beta\alpha^2), Tr(\beta\alpha^3)) \\ &= (Tr(\alpha^3), Tr(\alpha^4), Tr(\alpha^5), Tr(\alpha^6)) \\ &= (1, 0, 0, 1). \end{aligned}$$

7   Which Extractor Can Be Computed Efficiently?

The initial state $(s_0, s_1, \cdots, s_{n-1})$ is given by $s_t = Tr(\beta\alpha^t), t = 0, 1, \cdots, n-1$. From Sections 4-5, we know that we can select any k with $\gcd(k, 2^n - 1) = 1$ and $A_k \neq 0$ to solve for β, to obtain the initial state. The computation of the extractor is to compute A_k and $p_k(\alpha^k)$. For any k with $\gcd(k, 2^n - 1) = 1$, the cost for computing $p_k(x)$ are almost the same. Thus the difference between different k only depends on the computational cost of A_k.

If there, for example, is a term of degree d in the Boolean function and the corresponding tap positions are equally spaced, i.e., leading to a term $y_t = s_{t+i_0} s_{t+i_0+j} \cdots s_{t+i_0+(d-1)j}$ then for $k = 2^d - 1$, $A_k \neq 0$ (see [7]). In this case, let $\alpha_i = \alpha^{2^i j}, i = 0, 1, \cdots, d-1$. From Theorems 1-2, since there is only one solution for u, we have

$$A_{2^d-1} = \det E_u = \alpha^{i_0 (2^d - 1)} \det \begin{bmatrix} 1 & 1 & \cdots & 1 \\ \alpha_0 & \alpha_1 & \cdots & \alpha_{d-1} \\ \alpha_0^2 & \alpha_1^2 & \cdots & \alpha_{d-1}^2 \\ \vdots & \vdots & & \vdots \\ \alpha_0^{d-1} & \alpha_1^{d-1} & \cdots & \alpha_{d-1}^{d-1} \end{bmatrix} = \alpha_0^{d-1} \prod_{0 \le i < j < d} (\alpha_j + \alpha_i)$$

the last equality is due to fact that the matrix is a Vandermonde matrix. This is the simplest case for computing A_k. The computation complexity is the cost to compute $\binom{d}{2} = \frac{d(d-1)}{2}$ multiplications of two elements in $\mathbb{F}_{2^n}$, which is about $O(d^2 n^2)$ binary multiplications. In general, if the tap positions are not equally spaced, as long as there is one k with $H(k) = d$ such that $A_k \neq 0$. In this case, there is only one solution for u in Theorem 1. Thus $A_k = \det E_u$ (see Remark in Section 6). Therefore the computation of A_k for such a k is to compute the determinant of an $d \times d$ matrix E_u whose entries are taken from $\mathbb{F}_{2^n}$. This is the simplest case among all the other k's, since for the other k's where $H(k) < d$, there may be more than one solution for u in Theorem 1.

There is a trade-off between the number of known key stream bits of $\mathbf{b} = \{b_t\}$ and the computation cost in the pre-computation stage for forming an extractor. If we know $D = |\Omega(d)|$ consecutive bits of $\mathbf{b}$, then we do not need to compute the minimal polynomial of $\mathbf{b}$. Thus, we could select k wisely to reduce the cost for this particular A_k in pre-computation. If we only know L consecutive bits where L is the linear span of $\mathbf{b}$, then we have to compute the minimal polynomial of $\mathbf{b}$, which can be done by computing spectra $\{A_k\}$ by the method of Theorem 1 and Theorem 2. In this case, there is no saving for choosing special k, since we need to compute A_k for all $k \in \Gamma(d)$. We summarise the linear subspace attack, described in Sections 3-6, with these two different approaches in the pre-computation stage in the following two procedures.

Summary of the Linear Subspace Attack:
Procedure 1: $D = |\Omega(d)|$
Input: $b_0, \cdots, b_{D-1}$
Output: $s_0, \cdots, s_{n-1}$

1. Pre-computation:
   - For k with $\gcd(k, 2^n - 1) = 1$ and $H(k) = d$, compute A_k, and select k such that $A_k \neq 0$.
   - Compute $p_k(x) = \prod_{j \neq k, j \in \Gamma(d)} g_{\alpha^j}(x)$.
2. Compute the first n bits of $p_k(E)\mathbf{b}$, which is $u = (u_0, \cdots, u_{n-1})$.
3. Compute $x_0 = \frac{\det M_1}{\det M}$ where M is a Vandemonde matrix given by $\alpha_i = \alpha^{k2^i}, i = 0, 1, \cdots, n - 1$ (see Section 5) and M_1 is the matrix obtained by replacing the first column of M by u^T.
4. Solve for β from $x_0 = \beta^k A_k p_k(\alpha^k)$.
5. Compute $s_t = Tr(\beta\alpha^t), t = 0, \cdots, n - 1$.
6. Return $s_0, \cdots, s_{n-1}$.

Procedure 2: L, the linear span of b
Input: $b_0, \cdots, b_{L-1}$
Output: $s_0, \cdots, s_{n-1}$

1. Pre-computation:
   - For $j \in \Gamma(d)$, compute A_j and $g_{\alpha^j}(x)$ if $A_j \neq 0$.
   - Randomly select k with $\gcd(k, 2^n - 1) = 1$, and compute the polynomial
     $p_k(x) = \prod_{j \neq k: A_j \neq 0, j \in \Gamma(d)} g_{\alpha^j}(x)$.
The steps 2-6 are the same as in Procedure 1.

8   Discussion

If we do not know D (or L) consecutive bits of $\{b_t\}$, then, consequently, we do not have n consecutive bits of $\{u_t\}$. However, if we know $(u_{k_0}, u_{k_1}, \cdots, u_{k_{n-1}})$ from some known segments of $\mathbf{b}$, then the matrix M in Section 5 becomes $M = (m_{ij})$ where $m_{ij} = \alpha^{k_i 2^j}$ which may not be a Vandermonde matrix. In order to have the linear subspace attack to work when $det M \neq 0$, we can have an extractor for retrieving β, and therefore, the initial state of $\mathbf{s}$. So, the problem becomes that of how many bits we actually need to form an extractor with nonzero determinant of M. The method developed here can be also be applied to the case of combinatorial generators. We will discuss it in a separate paper.

Acknowledgements

This work was supported by the Norwegian Research Council.

References

1. Golomb, S.W., Gong, G.: Signal Design with Good Correlation: For Wireless Communications, Cryptography and Radar Applications. Cambridge University Press, Cambridge (2005)
2. Gong, G.: Analysis and Synthesis of Phases and Linear Complexity of Non-Linear Feedforward Sequences., Ph.D. thesis, University of Elec. Sci. and Tech. of China (1990)
3. Hawkes, P., Rose, G.: Rewriting variables: The complexity of fast algebraic attacks on stream ciphers. In: Franklin, M. (ed.) CRYPTO 2004. LNCS, vol. 3152, pp. 390–406. Springer, Heidelberg (2004)
4. Herlestam, T.: On Functions of Linear Shift Register Sequences. In: Pichler, F. (ed.) EUROCRYPT 1985. LNCS, vol. 219, pp. 119–129. Springer, Heidelberg (1986)
5. Paterson, K.G.: Root Counting, the DFT and the Linear Complexity of Nonlinear Filtering. Designs, Codes and Cryptography 14, 247–259 (1988)
6. Rønjom, S., Helleseth, T.: A New Attack on the Filter Generator. IEEE Transactions on Information Theory 53(5), 1752–1758 (2007)
7. Rueppel, R.A.: Analysis and Design of Stream Ciphers. Springer, Heidelberg (1986)
8. Rønjom, S., Helleseth, T.: The Linear Vector Space Spanned by the Nonlinear Filter Generator. In: Golomb, S.W., Gong, G., Helleseth, T., Song, H.-Y. (eds.) SEQUENCES 2007. LNCS, vol. 4893, Springer, Heidelberg (2007)

Author Index

Lecture Notes in Computer Science

Sublibrary 4: Security and Cryptology

Vol. 4301: D. Pointcheval, Y. Mu, K. Chen (Eds.), Cryptology and Network Security. XIII, 381 pages. 2006.

Vol. 4300: Y.Q. Shi (Ed.), Transactions on Data Hiding and Multimedia Security I. IX, 139 pages. 2006.

Vol. 4298: J.K. Lee, O. Yi, M. Yung (Eds.), Information Security Applications. XIV, 406 pages. 2007.

Vol. 4296: M.S. Rhee, B. Lee (Eds.), Information Security and Cryptology – ICISC 2006. XIII, 358 pages. 2006.

Vol. 4284: X. Lai, K. Chen (Eds.), Advances in Cryptology – ASIACRYPT 2006. XIV, 468 pages. 2006.

Vol. 4283: Y.Q. Shi, B. Jeon (Eds.), Digital Watermarking. XII, 474 pages. 2006.

Vol. 4266: H. Yoshiura, K. Sakurai, K. Rannenberg, Y. Murayama, S.-i. Kawamura (Eds.), Advances in Information and Computer Security. XIII, 438 pages. 2006.

Vol. 4258: G. Danezis, P. Golle (Eds.), Privacy Enhancing Technologies. VIII, 431 pages. 2006.

Vol. 4249: L. Goubin, M. Matsui (Eds.), Cryptographic Hardware and Embedded Systems - CHES 2006. XII, 462 pages. 2006.

Vol. 4237: H. Leitold, E.P. Markatos (Eds.), Communications and Multimedia Security. XII, 253 pages. 2006.

Vol. 4236: L. Breveglieri, I. Koren, D. Naccache, J.-P. Seifert (Eds.), Fault Diagnosis and Tolerance in Cryptography. XIII, 253 pages. 2006.

Vol. 4219: D. Zamboni, C. Krügel (Eds.), Recent Advances in Intrusion Detection. XII, 331 pages. 2006.

Vol. 4189: D. Gollmann, J. Meier, A. Sabelfeld (Eds.), Computer Security – ESORICS 2006. XI, 548 pages. 2006.

Vol. 4176: S.K. Katsikas, J. López, M. Backes, S. Gritzalis, B. Preneel (Eds.), Information Security. XIV, 548 pages. 2006.

Vol. 4117: C. Dwork (Ed.), Advances in Cryptology - CRYPTO 2006. XIII, 621 pages. 2006.

Vol. 4116: R. De Prisco, M. Yung (Eds.), Security and Cryptography for Networks. XI, 366 pages. 2006.

Vol. 4107: G. Di Crescenzo, A. Rubin (Eds.), Financial Cryptography and Data Security. XI, 327 pages. 2006.

Vol. 4083: S. Fischer-Hübner, S. Furnell, C. Lambrinoudakis (Eds.), Trust and Privacy in Digital Business. XIII, 243 pages. 2006.

Vol. 4064: R. Büschkes, P. Laskov (Eds.), Detection of Intrusions and Malware & Vulnerability Assessment. X, 195 pages. 2006.

Vol. 4058: L.M. Batten, R. Safavi-Naini (Eds.), Information Security and Privacy. XII, 446 pages. 2006.

Vol. 4047: M.J.B. Robshaw (Ed.), Fast Software Encryption. XI, 434 pages. 2006.

Vol. 4043: A.S. Atzeni, A. Lioy (Eds.), Public Key Infrastructure. XI, 261 pages. 2006.

Vol. 4004: S. Vaudenay (Ed.), Advances in Cryptology - EUROCRYPT 2006. XIV, 613 pages. 2006.

Vol. 3995: G. Müller (Ed.), Emerging Trends in Information and Communication Security. XX, 524 pages. 2006.

Vol. 3989: J. Zhou, M. Yung, F. Bao (Eds.), Applied Cryptography and Network Security. XIV, 488 pages. 2006.

Vol. 3969: Ø. Ytrehus (Ed.), Coding and Cryptography. XI, 443 pages. 2006.

Vol. 3958: M. Yung, Y. Dodis, A. Kiayias, T.G. Malkin (Eds.), Public Key Cryptography - PKC 2006. XIV, 543 pages. 2006.

Vol. 3957: B. Christianson, B. Crispo, J.A. Malcolm, M. Roe (Eds.), Security Protocols. IX, 325 pages. 2006.

Vol. 3956: G. Barthe, B. Grégoire, M. Huisman, J.-L. Lanet (Eds.), Construction and Analysis of Safe, Secure, and Interoperable Smart Devices. IX, 175 pages. 2006.

Vol. 3935: D.H. Won, S. Kim (Eds.), Information Security and Cryptology - ICISC 2005. XIV, 458 pages. 2006.

Vol. 3934: J.A. Clark, R.F. Paige, F.A.C. Polack, P.J. Brooke (Eds.), Security in Pervasive Computing. X, 243 pages. 2006.

Vol. 3928: J. Domingo-Ferrer, J. Posegga, D. Schreckling (Eds.), Smart Card Research and Advanced Applications. XI, 359 pages. 2006.

Vol. 3919: R. Safavi-Naini, M. Yung (Eds.), Digital Rights Management. XI, 357 pages. 2006.

Vol. 3903: K. Chen, R. Deng, X. Lai, J. Zhou (Eds.), Information Security Practice and Experience. XIV, 392 pages. 2006.

Vol. 3897: B. Preneel, S. Tavares (Eds.), Selected Areas in Cryptography. XI, 371 pages. 2006.

Vol. 3876: S. Halevi, T. Rabin (Eds.), Theory of Cryptography. XI, 617 pages. 2006.

Vol. 3866: T. Dimitrakos, F. Martinelli, P.Y.A. Ryan, S. Schneider (Eds.), Formal Aspects in Security and Trust. X, 259 pages. 2006.

Vol. 3860: D. Pointcheval (Ed.), Topics in Cryptology – CT-RSA 2006. XI, 365 pages. 2006.

Vol. 3858: A. Valdes, D. Zamboni (Eds.), Recent Advances in Intrusion Detection. X, 351 pages. 2006.

Vol. 3856: G. Danezis, D. Martin (Eds.), Privacy Enhancing Technologies. VIII, 273 pages. 2006.

Vol. 3786: J.-S. Song, T. Kwon, M. Yung (Eds.), Information Security Applications. XI, 378 pages. 2006.

Vol. 3108: H. Wang, J. Pieprzyk, V. Varadharajan (Eds.), Information Security and Privacy. XII, 494 pages. 2004.

Vol. 2951: M. Naor (Ed.), Theory of Cryptography. XI, 523 pages. 2004.

Vol. 2742: R.N. Wright (Ed.), Financial Cryptography. VIII, 321 pages. 2003.